Data Science and SDGs

Bikas Kumar Sinha · Md. Nurul Haque Mollah
Editors

Data Science and SDGs

Challenges, Opportunities and Realities

Springer

Editors
Bikas Kumar Sinha
Indian Statistical Institute
Kolkata, West Bengal, India

Md. Nurul Haque Mollah
Department of Statistics
University of Rajshahi
Rajshahi, Bangladesh

ISBN 978-981-16-1918-2 ISBN 978-981-16-1919-9 (eBook)
https://doi.org/10.1007/978-981-16-1919-9

This Springer imprint is published by the registered company Springer Nature Singapore Pte Ltd.
The registered company address is: 152 Beach Road, #21-01/04 Gateway East, Singapore 189721, Singapore

Foreword

Many of us are presumably aware of the Govt. of Bangladesh Report (2017) on "Data Gap Analysis for Sustainable Development Goals (SDGs): Bangladesh Perspective" by the Planning Commission, Govt. of Bangladesh, General Economics Division (GED) (Making Growth Work for the Poor) Planning Commission, January 2017. Keeping this background in mind, the Department of Statistics, University of Rajshahi, organized an appropriate and timely "International Conference on Data Science and SDGs: Challenges, Opportunities and Realities" held in December 18–19, 2019.

As a Chairperson of both the Opening and Closing Sessions of the conference, I was indeed delighted to witness the success of the events that brought together so many stakeholders from divers fields, and also the tremendous interest it generated among teachers, research scholars and student alike.

The conference was attended by over 300 participants from different countries including Bangladesh, Finland, India, Korea, Malaysia and USA. There were 150 Speakers (2 Keynote, 5 Plenary, 36 Invited and 107 Contributed) who presented their papers; there were 60 posters and 27 organizations/research groups took part in the exhibition. It is a matter of great pleasure that the conference was also attended by policymakers, researchers, government entities, and nongovernment organizations.

The organizers of the conference did not waste any time to visualize the importance of publication of Conference Proceedings and the approach was made to Springer Nature (SN), New Dehli with necessary documents. The proposal was readily accepted by SN. I thank the organizers and the Editorial Board members for their commendable effort.

The proceedings volume covers topic mainly on SDGs, bioinformatics, public health, medical informatics, environmental statistics, data science and machine learning. It is understood that almost all of the topics are covered by thoroughly statistical models and methods, with examples using real data including graphical illustrations, as and wherever necessary.

The proposed book can be used as a research monograph for practitioners or as a reference text for the researchers. As a research monograph, it unifies the literature on data science and SDGs and parents important result in an integrated manner and highlights topics that need further research. It is also suitable for use as a reference text for graduate and advanced undergraduate level programs in SDGs, bioinformatics, public health, medical informatics, environmental statistics, data science and machine learning.

I hope that this volume will be very useful for ushering analytical insights in data science and SDGs in communities around the world. I am very happy to endorse this publication anticipating that it will foster further research interest in attaining the SDGs.

Prof. M. Abdus Sobhan
Vice Chancellor
University of Rajshahi
Rajshahi, Bangladesh

Preface

To a large extent, 17 Sustainable Development Goals (SDGs), also known as the urgent Global Goals, were adopted by all United Nations Member States in 2015 as a universal 'call to action' to end poverty, protect the planet and ensure that all people enjoy peace and prosperity by 2030.

Monitoring the progress made by different countries towards meeting the SDGs is extremely crucial to track progress. Towards this, concerted efforts of the Government of Bangladesh are recorded in a Report (2017). Data-driven measurement of progress needs to be distributed to stakeholders, mainly policymakers, researchers, government entities, civil society, nonprofit organizations, etc. Decision based on data can be used to re-align and influence governments, businesses, citizens to accelerate the progress. Measuring and controlling data on SDGs is quite difficult to handle and distribute by using traditional data gathering and manipulation techniques due to the volume and distribution of data points around the globe. This is where Data Science, especially the Big Data issues come into play.

Bringing together these stakeholders was the goal of the 'International Conference on Data Science and SDGs: Challenges, Opportunities and Realities' organized by the Department of Statistics, University of Rajshahi, Bangladesh during December 18–19, 2019. This Department tries to organize an international conference in three years' interval on different burning issues, and it was the 7th of such events. The conference was attended by over 300 participants from different countries including Bangladesh, Finland, India, Korea, Nepal, Malaysia and USA. There were 150 Speakers (2 Keynote, 5 Plenary, 36 Invited and 107 Contributed) who presented their papers; 60 presenters presented their posters and 27 organizations/research groups took part in the exhibition.

Immediately after the conclusion of the conference, the organizing committee and the programme committee decided to publish proceedings volumes. Accordingly, an Editorial Board was formed with ten (10) members, keeping Springer Nature (SN), New Delhi in mind:

Professors (1) Bikas K Sinha (Retired Professor, Indian Statistical Institute, Kolkata) and (2) Md. Nurul Haque Mollah (University of Rajshahi)—Joint Co-ordinators

(3) Prof. Malay Ghosh (University of Florida, Gainesville, USA)

(4) Prof. Tapio Nummi (Tampere University, Finland)

(5) Prof. Rahmatullah Imon (Ball State University, USA)

(6) Prof. Jyotirmoy Sarkar (Indian University—Purdue University, Indianapolis, USA)

Professors (7) M. Sayedur Rahman, (8) Md. Ayub Ali, (9) Md. Golam Hossain and (10) Md. Rezaul Karim—all of the Department of Statistics, University of Rajshahi.

The Joint Co-ordinators got in touch with Springer Nature (SN) New Delhi Publishers and sent the Book Proposal with all necessary documents. The proposal included a list of 15 Chapters based on personal presentations by the participants in different categories. The papers were reviewed by a group of subject experts from India (9) and USA (6). The proposal was readily accepted by SN, and we entered into an agreement. We thank all the contributors and the reviewers for enriching this publication of the Conference Proceedings Volume. We sincerely thank the learned reviewers for rendering their intellectual service towards accomplishing our goals. We list their names below.

(1) Dr G. C. Manna, Ex-DG, CSO, GoI, Ex-Member, National Statistical Commission, GoI

(2) Dr. R. R. Nandy, Associate Professor, Epidemiology & Bio-Statistics, Univ. North Texas, USA

(3) Prof. Sati Majumdar, Distinguished Bio-Statistician, Medical School, University of Pittsburgh, USA

(4) Dr. Avinash Dharmadhikari, Ex-GM, Tata Motors, Pune

(5) Prof. Aditya Bagchi, Emeritus Professor, Computer Science Dept. School of Mathematical Sciences, Ramakrishna Mission Vivekananda Educational and Research Institute, Kolkata

(6) Prof. Gopal Basak, Stat-Math Division, Indian Statistical Institute, Kolkata

(7) Prof. Dulal K. Bhaumik, Bio-Statistician in Medical School, Univ. Illinois, Chicago (UIC), USA

(8) Prof. Premananda Bharati, Anthropology & Human Genetics Unit, Indian Statistical Institute, Kolkata

(9) Dr. Moumita Chatterjee, Department of Statistics, Aliah University, Kolkata

(10) Prof. Parasmani Dasgupta, Biological Anthropology Unit, Indian Statistical Institute, Kolkata

(11) Prof. Nripes K. Mandal, Department of Statistics, Calcutta University, Kolkata

Besides this publication, there are 5 *International Journal of Statistical Science* (*IJSS*) regular issues [2019, 2020(1), 2020(2) 2021(1), 2021(2)]—incorporating a total of around 45 manuscripts, based on technical presentations made by the presenters during the conference and further improvement according to the reviewer comments. These regular issues have been/are being processed by the IJSS EB.

In organizing such a mega-event and the follow-up journal publications, financial support has come from around 50 agencies including Bangladesh Bank (BB), Northwest power generation (NPG), Investment Corporation Bangladesh (ICB), Islami Bank Bangladesh Limited (IBBL), Rajshahi Krishi Unnayan Bank (RAKUB), Bangladesh Investment Development Authority (BIDA), Bangladesh Bureau of Statistics (BBS) and Bangladesh Rice Research Institute (BRRI), and we, on behalf of the EB Members of this Proc. Volume, offer our sincere thanks to them.

The Department of Statistics, University of Rajshahi, is the main architect of this mega-event, followed by the on-going publications. We thank the HoD and other Faculty and Non-Teaching Staff of the Department for rendering such opportunities to us. Needless to say, IJSS Office-bearers are shouldering enormous amount of responsibilities during the process of publication of the proceedings volumes. We thank them individually and collectively. We also would like to acknowledge Dr. Md. Abul Basher Mian, retired professor of Statistics Department of Rajshahi University (RU) and Dr. Rehman Eon, Research Scientist, Rochester Institute of Technology (RIT), Rochester, USA, for grammatical checking of some manuscripts.

Lastly, we thank Ms. Nupoor Singh of SN, New Delhi, for handling our project proposal with extreme courtesy at all stages.

We will consider our efforts amply rewarded if this publication creates enough interest among the stakeholders at large, including researches in the areas of Data Science and Big Data Analytics.

Kolkata, India | Bikas Kumar Sinha
Rajshahi, Bangladesh | Md. Nurul Haque Mollah

Pictures Conference

Second Floor, Department of Statistics
Sir Jagadish Chandra Bose Science Building

Poster of International Conference 2019

Audience, International Conference 2019
Shahid Tajuddin Ahmad Senate Bhaban, Date: 18-11-2019

Exhibition Visit by the Honorable Chief Coordinator of SDGs, Prime Minister's Office International Conference 2019, Date: 18-11-2019

Exhibition Visit by the Honorable Chief Coordinator of SDGs, Prime Minister's Office International Conference 2019, Date: 18-11-2019

A Group of Scientist
International Conference 2019, Date: 19-11-2019

Exhibition Visit by the Honorable Governor, Bangladesh Bank
International Conference 2019, Date: 19-11-2019

Guests of Cultural Evening, Kazi Nazrul Islam Auditorium
International Conference 2019, Date: 19-11-2019

Contents

Editors and Contributors

About the Editors

Prof. Bikas Kumar Sinha has been publishing Springer monographs and edited volumes since 1989. Besides being an academic with Indian Statistical Institute [1979-2011], he has served Government of India as a Member of National Statistical Commission for the 3-year period 2006-2009. His research publications span both statistical theory and applications. He was appointed an 'Expert on Mission' for the United Nations-funded Workshop on Survey Methodology in 1991. Professor Sinha has travelled extensively worldwide with teaching/research/consultancy assignments. He has held various academic assignments in international institutes.

Prof. Md. Nurul Haque Mollah is attached to the Department of Statistics, University of Rajshahi, Bangladesh. He completed his PhD degree research (2002-2005) in Statistical Science from the Institute of Statistical Mathematics (ISM), Tokyo, Japan. He also completed his post-doctoral research (2006-2008) in bioinformatics from ISM, Tokyo, Japan. His area of research specialization covers multivariate statistics, robust statistical inference, statistical signal processing, data mining, bioinformatics and computational biology. He has published 170 research articles, book chapters in refereed journals, books and proceedings of international conferences so far. He has supervised more than 50 research students at Master and PhD levels. He successfully led, as the Principal Investigator (PI), several research projects funded by UGC, NST, BANBEIS and World Bank. He also organized several national/international workshops, seminars and conferences as convener/co-convener. He is the convener of BioRGSD (Bioinformatics Research Group of Statistics Department, RU) & BioRGRU (Bioinformatics Research Group, RU). He is also the Founder President of BBCBA (Bangladesh Bioinformatics and Computational Biology Association).

Contributors

Shamsul Alam General Economics Division, Bangladesh Planning Commission, Dhaka, Bangladesh

Md. Ayub Ali Department of Statistics, University of Rajshahi, Rajshahi, Bangladesh

Md. Borqat Ali Department of Statistics, University of Rajshahi, Rajshahi, Bangladesh

Abu Sayed Md. Al Mamun Department of Statistics, University of Rajshahi, Rajshahi, Bangladesh

Md. Asadul Alam Department of Mathematics, Daud Public College, Jessore, Bangladesh

Abdul Aziz Department of CSE, Khulna University of Engineering & Technology, Khulna, Bangladesh

Priyanka Bosu Department of Statistics, Faculty of Science, Bangabandhu Sheikh Mujibur Rahman Science and Technology University, Gopalganj, Bangladesh

Md. Abeed Hossain Chowdhury BARC, Dhaka, Bangladesh

Biswa Nath Datta Department of Mathematical Sciences, Northern Illinois University, De Kalb, IL, USA;
Department of Mathematics, Indian Institute of Technology, Kharagpur, India

Md. Manzur Rahman Farazi Medical College of Wisconsin, Shorewood, WI, USA

Mst. Farzana Akter Department of Statistics, Shahjalal University of Science and Technology, Sylhet, Bangladesh

Rowshanul Habib Department of Biochemistry and Molecular Biology, University of Rajshahi, Rajshahi-6205, Bangladesh

Md. Harun-Or-Roshid Department of Statistics, University of Rajshahi, Rajshahi, Bangladesh

Md. Golam Hossain Department of Statistics, University of Rajshahi, Rajshahi, Bangladesh

Md. Kamrul Islam Department of Statistics, University of Rajshahi, Rajshahi, Bangladesh

Md. Moidul Islam Department of Statistics, University of Rajshahi, Rajshahi, Bangladesh

Md. Saiful Islam Department of Biochemistry and Molecular Biology, University of Rajshahi, Rajshahi-6205, Bangladesh

A. R. M. Jalal Uddin Jamali Department of Mathematics, Khulna University of Engineering & Technology, Khulna, Bangladesh

Jesmin Department of Genetic Engineering and Biotechnology, University of Dhaka, Dhaka, Bangladesh

Md. Rezaul Karim Department of Applied Nutrition and Food Technology, Islamic University, Kushtia-7003, Bangladesh

Md. Abdul Khalek Department of Statistics, University of Rajshahi, Rajshahi, Bangladesh

Suman Khan Department of Statistics, Faculty of Science, Bangabandhu Sheikh Mujibur Rahman Science and Technology University, Gopalganj, Bangladesh

Md. Akhtaruzzaman Limon Department of Statistics, University of Rajshahi, Rajshahi, Bangladesh

Md. Nurul Haque Mollah Department of Statistics, University of Rajshahi, Rajshahi, Bangladesh

Md. Ashek Al Naim Department of Statistics, University of Rajshahi, Rajshahi, Bangladesh

Tapio Nummi Faculty of Information Technology and Communication Sciences, Tampere University, Tampere, Finland

Mohammad Ohid Ullah Department of Statistics, Shahjalal University of Science and Technology, Sylhet, Bangladesh

Md. Matiur Rahaman Department of Statistics, Faculty of Science, Bangabandhu Sheikh Mujibur Rahman Science and Technology University, Gopalganj, Bangladesh

M. Sayedur Rahman Environment and Data Mining Research Group, Department of Statistics, University of Rajshahi, Rajshahi, Bangladesh

Md. Mostafizur Rahman Environment and Data Mining Research Group, Department of Statistics, University of Rajshahi, Rajshahi, Bangladesh

A. H. M. Rahmatullah Imon Department of Mathematical Sciences, Ball State University, Muncie, IN, USA

Md. Razanmiah Department of Statistics, University of Rajshahi, Rajshahi, Bangladesh

Biswajit Sahoo Department of Mechanical Engineering, Indian Institute of Technology, Kharagpur, West Bengal, India

Bandhan Sarker Department of Statistics, Faculty of Science, Bangabandhu Sheikh Mujibur Rahman Science and Technology University, Gopalganj, Bangladesh

Bikas K. Sinha Indian Statistical Institute, Kolkata, India

Kumkum Yeasmin Department of Statistics, University of Rajshahi, Rajshahi, Bangladesh

Tanzima Yeasmin Department of Biochemistry and Molecular Biology, University of Rajshahi, Rajshahi-6205, Bangladesh

SDGs in Bangladesh: Implementation Challenges and Way Forward

Shamsul Alam

Sustainable Development Goals (SDGs) are adopted as post-2015 development agenda by UN, which comprise of 17 goals and 169 targets. This agenda calls for action by all countries; poor, rich, and middle income. For its proper implementation, General Economics Division (GED) of Bangladesh Planning Commission has devised mapping of ministries (2016) by targets and formulated national action plan (2018) related to SDG goals and targets linking SDG targets with the current 7th FYP for harmonizing national and global agenda. However, the daunting task of gathering indicator-wise data for monitoring SDG progress remains a challenge. Only data for one-third of total indicators are available in Bangladesh. On the other hand, SDGs implementation requires a huge amount of resources for the remaining period, and a recent study by GED (2018) suggests five potential sources of gap financing, which culminates into a "Whole of Society" approach. This requires highest level of cooperation among the stakeholders and world communities. Strong diplomatic ties and enhanced bureaucratic and management capacity are expected to play significant role in achieving the SDG targets through necessary resource mobilization, especially in the post-LDC graduation era for Bangladesh.

This chapter draws some material from the author's previous work published in the annual report of General Economics Division (GED), Bangladesh Planning Commission, Bangladesh. The author has received necessary permission from the publisher of the said annual report.

S. Alam (✉)
General Economics Division, Bangladesh Planning Commission, Dhaka, Bangladesh
e-mail: member.ged@plancomm.gov.bd

B. K. Sinha et al. (eds.), *Data Science and SDGs*,
https://doi.org/10.1007/978-981-16-1919-9_1

1 SDGs and Bangladesh's Engagement

The United Nations General Assembly at the 70th session held on 25 September 2015 adopted the outcome document of the UN summit for the adoption of the post-2015 development agenda entitled "Transforming Our World: the 2030 Agenda for Sustainable Development" and decided on new global Sustainable Development goals (SDGs). At the core of the 2030 Agenda is a list of 17 Sustainable Development Goals (SDGs) and 169 targets to end poverty, hunger, inequality, take action on climate change and the environment, improve access to health and education, care for people and the planet, and build strong institutions and partnerships. The SDGs are unprecedented in terms of scope and significance and go much beyond the MDGs (2001–15) by including economic growth, sustainable production and consumption, sustainable urbanization, data generation for tracking progress, and the importance of peace and justice for all in the agenda adding new dimensions.

Bangladesh has gained worldwide recognition as one of the pioneers during MDGs implementation. Due to the strong leadership of Honorable Prime Minister Sheikh Hasina, Bangladesh attained many targets ahead of time. Some special areas where tremendous progress has been achieved are poverty, food security, enrolment at primary and secondary levels, lower mortality rates of infant and their mothers, immunization, etc. Because of its excellent achievements during MDG.

Under the guardianship of the General Economics Division (GED), Planning Commission of Bangladesh, the drafting of the post-2015 Development Agenda was initiated. GED, on behalf of the government of Bangladesh, also suggested several goals and associated targets along with indicators in this draft. The process followed the whole of society approach involving consultations with government, private, NGOs, CSOs, and academia at the national and local levels. In this draft document, Bangladesh proposed 11 goals, 58 targets, and 241 indicators among which nine goals were matched with the proposals of the Open Working Group of United Nations and two other goals were partially covered under the targets of related SDGs. SDGs are completely voluntary in nature. However, all the signatory countries are expected to undertake appropriate measures to prepare a national framework for the implementation of this global agenda.

2 Brief Note on Implementation of SDGs

- **Institutional Arrangements for Achieving SDGs**

Recognizing the challenge of implementation of ambitious and transformational SDGs, the Prime Minister established an Inter-Ministerial Committee on SDGs Monitoring and Implementation demonstrating her commitment to SDGs.

- **Mapping of Ministries by Sustainable Development Goals and Targets**

The 7th Five Year Plan comprises two broad parts: Part 1 focuses on macroeconomic perspective and Part 2 focuses on sector development strategies spreading over 13 sectors. The plan is implemented mainly through programs and projects derived from the sectoral strategies. The sectors are highly aggregative, and a number of Ministries/Divisions are responsible for preparing and implementing projects/programs under a particular sector. A cursory look at the targets of SDGs indicates a complex web of Ministries/Divisions are responsible for attaining a particular target. In order to delineate the responsibilities of different ministries/divisions to each of the targets, the Government has mapped the relevant ministries/divisions by goal and associated target. The mapping exercise has assigned the lead role in attaining a target to a ministry/division or organization, which is supported in most cases by a co-lead ministry/division.

- **Preparation of Ministry/Division Action Plan and the National Action Plan for the Implementation of SDGs**

As a sequel to the mapping exercise, the ministries/divisions/organizations are required to prepare their respective action plans which would have specific actions/activities and interventions to achieve their respective goals/targets. The Ministries consulted both the 2030 Agenda and the current FYP (7th) to formulate short, medium, and long-term sector specific plans for the next five years.

The National Action Plan for the implementation of the SDGs has been prepared by GED, Bangladesh Planning Commission, by consultation with the ministries/divisions. The Plan also lists the ongoing projects/programs that contribute to the achievement of a goal and its targets, and identifies new projects/program that will need to be undertaken during the remaining period of the 7th Plan as well as beyond the current Plan period with indicative costs. The Plan represents a dynamic/living document that leaves scope for amendment/revision during the implementation of the 8th Five Year Plan (2021–2025).

- **Data Gap Analysis for SDGs**

The need for data and statistics to monitor the progress on SDGs to ensure their full implementation by target date was amply demonstrated by the United Nations Secretary General's High Level Panel of Eminent Persons on the Post-2015 Development Agenda which called for a "data revolution" in 2013. It is also realized that countries especially developing countries do not currently generate all the data needed for monitoring the relevant indicators. Accordingly, Bangladesh undertook two separate exercises, one by the BBS and the other by the Planning Commission, to understand the current state of data and take informed policy decisions on the implementation of the post-2015 agenda.

The General Economics Division (2017) undertook an assessment of the current status of data in the country, the availability of data from different sources, and the

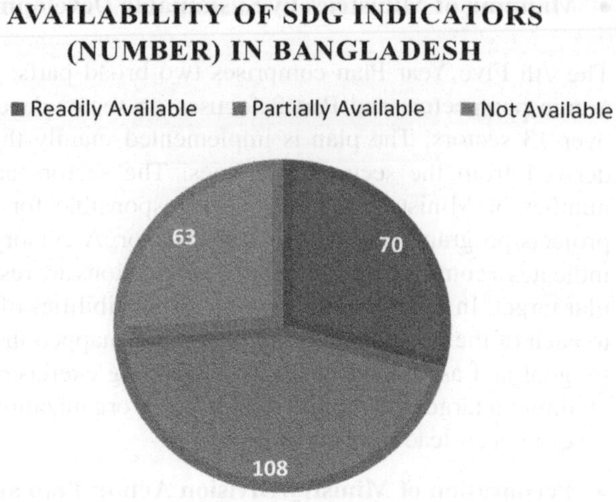

Fig. 1 Availability of data of SDG indicators in Bangladesh. *Source* Data gap analysis for SDGs: Bangladesh Perspective (2017), GED, Bangladesh

gap that needs to be filled through generation of new data. The exercise involved all the relevant data generating agencies including Bangladesh Bureau of Statistics.

The report divides the indicators into three types depending on the status of data availability: (i) indicators for which data are readily available, (ii) Indicators for which data are partially available, meaning that some modification, addition, and analysis are required in the existing census or survey for obtaining pertinent data, and (iii) Indicators for which data are not available giving rise to need for new census or survey. It is observed that 70 indicators (29%) belong to the first category, 63 indicators (26%) belong to the third category, and 108 indicators (45%) belong to the second category. Data availability, timeliness, and quality pose a significant challenge to effective monitoring that could help policy decisions (Fig. 1).

- **Monitoring and Evaluation Framework of SDGs**

The Monitoring and Evaluation Framework of SDGs (GED, 2018) has been developed to track progress on implementation and achievement of SDGs in Bangladesh in the next 13 years. Several issues need to be highlighted before discussing the M&E framework. First, because of the wide range of aspects of the economy and its depth that need to be measured to assess the progress on SDGs, the set of indicators to measure progress is diverse and complex. In many cases a target is not measured by a single number rather by multiple numbers depending on the level of disaggregation. Second, BBS does not generate data on many aspects of the economy to meet the requirements of our national development plans. Consequently, data on many indicators are simply not available. Third, data are generated by BBS or other government agencies through periodic surveys. The periodicity varies from five years for Household Income and Expenditure Survey to three years for Bangladesh Demographic and Health Survey. Interestingly, BBS has been conducting Labour Force Survey at three-year intervals. Fourth, data generation through more frequent surveys as well as generation of more disaggregated data (such as spatial, gender, age-group, ethnicity, employment status) will require increasing financial and human resources, logistics support as well as use of modern technology.

The monitoring framework provides baseline data of each indicator, which are available for the targeted terminal year of SDGs, i.e., 2030 with two milestones at 2020 and 2025. The data source in the document consists of the information regarding the concerned ministries/divisions and their respective agencies or departments that generate the data for an indicator. Efforts are underway to revise/reassess the Monitoring and Evaluation Framework on the 4th year of SDGs implementation with the availability of the latest SDGs indicators suggested by IAEA–SDGs in 2020.

- **SDGs Needs Assessment and Financing Strategy**

Implementation of ambitious SDGs requires huge amount of resources during the 2016–2030 period. Mobilization and effective use of this resource pose considerable challenge to the developing countries. Bangladesh is committed to achieve SDGs and hence needs to estimate the amount of resources that will be required, financing sources, instruments, and strategies. The "SDGs Financing Strategy: Bangladesh Perspective" prepared by the General Economics Division (GED) of the Planning Commission provides an estimate of the annual resource gap. The estimates show that an additional amount, over the current provision of investment related to SDGs by public sectors and external sources, would be $928.48 billion at 2015–16 constant prices spreading over a period of 13 years (2018–2030), which is 19.75% of the accumulated gross domestic product (GDP). The annual average cost of SDGs would be $66.32 billion (at constant prices) for this period. The costing exercise covers close to 80% of the 169 targets of SDGs because of inter-connectivity of targets to each other. This helps to avoid over-estimation of costs. Implementation of SDGs globally would require $3–5 trillion annually as predicted by the UN.

The study has suggested five potential sources of gap financing. These are Private Sector Financing, Public Sector Financing, Public–Private Partnership (PPP), External Financing comprising Foreign Direct Investment (FDI) and foreign aid and grants, and Non-Governmental Organizations (NGOs). On an average, public sector would account for around 34% of the financing requirement, whereas private sector has the share of around 42% during 2018–2030 period. The Goals and associated targets of SDGs have large public goods aspect whose provision would require higher public funding relative to private sector's contribution. The average share of PPP is 6.0%. The external sources would constitute close to 15% where FDI would make up 10% and foreign aid would comprise 5.0% of financing gap. Finally, the NGOs would contribute around 4.0% for the same period. Despite high expectations, the status of additional fund mobilization for the implementation of SDGs fell short of set targets. Approximately BDT 3772 billion additional funding is required for implementation of SDGs from FY 2017 to FY 2020 (source: GED) against which the fiscal budget increased by BDT 2280 billion. Therefore, there exists a visible gap between public sector's estimated and actual contribution towards SDGs. Nevertheless, the coming Five Year Plan of the government, which is scheduled to be kicked off from July 2020, will align SDGs with increased budgetary allocation and will try to draw budgetary supports from multilateral lending agencies that will ensure successful implementation of 2030 Development Agenda.

3 The "Whole of Society" Approach and Coordination Among Different Stakeholders in the Implementation of SDGs in Bangladesh

The Government of Bangladesh has consistently applied "whole of society" approach to the preparation of national development plans and policy documents of national importance.

In view of the critical role of the private sector in attaining SDGs, consultation meetings between the Government of Bangladesh, private sector, and the UN System on the "*Role of the Private Sector in Facilitating the SDGs*" have been held to highlight the broad outlines for private sector actions on SDGs implementation. This has been done in addition to several rounds of consultations held with civil society organizations and NGOs. The Government also appreciates the value of media in creating awareness of the people. Effective and coherent role of both print and electronic media in creating SDGs awareness and branding of success would be strongly needed. GED held meetings with the media and produced 21 SDGs related documents in creating broader based public awareness.

4 Successes Achieved So Far in SDGs Implementation in Bangladesh

- Bangladesh has made remarkable progress in its effort on reducing extreme poverty. In 2019, 10.5% of the population was living below the international poverty line of $1.90/day, which was 19.6% in 2010. Similarly, 20.5% of the population was living below the national poverty line in 2019; a considerable improvement than 2010 (31.5%). The expansion of social protection and public expenditure on basic services also gained a momentum in the last ten years. The social protection coverage climbed to 58.1% in 2019 as opposed to 24.6% in 2010.
- Despite the sub-regional regression, progress on reducing stunting which stood at 28.0% in 2019 is virtually on track at the current rate of reduction. Similarly, progress on reducing wasting which stood at 8.4% is also on track.
- Progress has also been attained in gender equality (SDG-5). The share of female parliamentarian is gradually increasing. The proportion was 20.86% in 2019 compared to 12.42% in 2001. Women participation in education and workforce also went up in the last few years, which is promising. The Global Gender Gap Report 2018 by the World Economic Forum ranked Bangladesh as the only country

in the sub-continent under 100 (48th position) in women empowerment. is the only country in South Asia which has been ranked under 100 (48th) in women empowerment.

Bangladesh is moving relentlessly towards ensuring access of 100% households to electricity. In the year 2019, 92.2% of the population had access to the electricity, which stood at a meager 55.26% in 2010, and it is expected that the rate would reach 97% by 2020. However, the share of population having access to clean fuel and modern technology for cooking stood at only 19%. With regard to renewable energy, government is making all-out efforts to increase the share of the renewable energy in the energy mix.

- The economy of Bangladesh has made a positive shift with regard to the growth rate of GDP. In recent years, the average annual growth rate of GDP has surpassed 7%. Along with the GDP growth, the average growth of GDP per capita has also shown optimism. For a reasonable period of time, Bangladesh has been maintaining a very stable unemployment rate, which roughly stands at around 4%. Recent GED Survey revealed that the unemployment rate in Bangladesh is 3.1% (2018).
- Total international support to infrastructure has been increasing with some annual fluctuation. The proportion of population covered by 2G mobile network has reached close to 100%, while the 2020 milestone has already been achieved in June 2019 in case of 3G technology. 4G coverage has reached to 79% in June 2019.
- Bangladesh has made a commendable progress in the area of environment and climate change. The country already approved the Disaster Risk Reduction Strategies of Bangladesh (2016–2020) in line with the Sendai Framework for Disaster Risk Reduction 2015–2030. During the period from 2014–2020, the number of deaths, missing persons, and directly affected persons due to natural disasters has gone down by half. With this rate of progress, Bangladesh is hopeful to meet the target within the stipulated timeframe.
- Recently, Bangladesh has secured a vast swath of marine territory (118,813 km^2 in total). The area is abundant in natural resources and possesses a rich bio-diversity. It is very important to conserve and explore these resources in a sustainable manner. Based on the statistics of the Department of Forest, the coverage of the protected areas in relation to the total marine areas was 7.94% in 2016–17.
- Numerous efforts have been undertaken by the government to achieve the SDG-15. Some of the efforts include declaration of Ecologically Critical Areas (ECAs), creation of bio-diversity zones, etc. Apart from strategies and plans, proper implementation to preserve the bio-diversity is an exigency. According to an estimate of the Bangladesh Forest Department, the forest coverage in Bangladesh was 14.47% in 2018. The government has initiated coastal afforestation program in its large swath of newly accreted coastal zone. Mangrove plantation is underway covering a gigantic 140,000 ha of land. Apart from these efforts, the government declared

21 Protected Areas for the conservation of wildlife. Among these protected areas, the share of terrestrial area has gone up from 1.7% in 2015 to 3.08% in 2018.

- Successful implementation and the attainment of SDGs depend on the availability of resources and global partnership. Government revenue as proportion of GDP has been increased more than the required rate in the recent period (11.60% in FY2017-18 compared to 9.6% in 2015) due to the measures undertaken for increasing the number of tax payers and improving tax collection and management mechanism. However, the ODA statistic indicates a sluggish growth despite the fact that ODA's share in the national budget is gradually decreasing. With regard to the FDI and remittance inflow, the government needs a quick boost for mobilizing necessary resources in the SDGs' implementation process. In this regard, Bangladesh government has explored the potential avenues for international cooperation. In a recent study by GED, the government found that 62 SDG targets need enhanced international support and deemed crucial for successful and timely implementation of the post-2015 development agenda.
- Government has made remarkable progress so far in minimizing SDGs data gap. During the first SDGs Monitoring and Evaluation Framework, baseline data were available for 127 indicators. In the revised framework, due to be published in June 2020, this figure rose to 161. Moreover, National Statistics Organization has identified 19 new surveys, among which 6 surveys have already been completed, 11 surveys are in the pipeline for being completed, and 2 surveys will be undertaken soon to update old data. Moreover, 2 censuses are underway, 2 projects are being implemented, and the revision of the National Statistics Act is ongoing. All these efforts are meant to make the data available for 44 new SDGs indicators and complement 59 other SDGs indicators for which data were partially available previously. The National Statistical Organization has more than thousand personnel who are working from the central administration to the upazila level, one of the lowest tiers of local government administration, and they have been performing well in managing all the surveys and censuses. On-the-job training and participation at national and global workshops have enriched their knowledge in carrying out the surveys and census with subtlety. However, more focus is required on gathering rich administrative data that will be essential for periodic monitoring of an indicator, as surveys are less frequent and involve much financial prowess of the government.

5 Revisiting Challenges of SDGs Implementation in Bangladesh

- **Complexity of Targets**

In many cases, the targets are complex and emphasize many different aspects. It may be said, SDGs cover too many targets at a time. Whole SDGs were seemingly done academically rather than considering practical aspects of implementation. Nonetheless, achievement of broad goals/targets will make our planet a much better place to live. Achieving a target requires investment that can contribute to improving some aspects of the target, but not all. In such cases, investment might have to be spread over several projects/programs that will ultimately increase the size of total outlay.

- **Synergy and Dissonance of SDGs**

Designing and implementing policies of SDGs at the end of 2030 poses considerable challenge for planners and policymakers. Many of the SDGs are linked, sometimes in subtle ways to others. SDG policy in one sector can also cause unanticipated underachievement in another sector. For example, if a nation (least developed or developing) emphasizes Goal 8 through private and public investment for accelerated growth for job creation, it may affect rise of inequality concerning Goal 10 as of in the early development process. Conversely, SDG policy in one sector can cause synergistic overachievement in another sector. Policies and programs to achieve SDGs need to consider these complex linkages characterized by synergy and dissonance.

- **Ranking in International Indices and Attainment of SDGs**

Ratings and rankings can be powerful tools of both branding and influencing citizenry. A big reason for the boom in indices is their growing use by governments, NGOs, and campaigners to shape new laws and getting them passed. Government's position on these rankings helps to capture national and international attention and deems important for any form of cooperation. Moreover, Global Performance Indicators should be thought of increasingly as tools of global governance.

Bangladesh's progression in global indices is important for several reasons: SDGs attainment, improved comparative status among countries, and mapping of weak areas for policy making are the most notable ones.

Due to the unavailability of important data, Bangladesh faces a great challenge in positioning themselves in global indices. The major challenges are:

- Lack of coordination among data providing agencies
- Lack of indicator-wise metadata
- Lack of data availability in proper format
- Lack of methodological understanding, etc.

The government of Bangladesh recently took an initiative to gather data for assisting in the construction of 23 international Indices which fall in the following thematic areas:

- Poverty, hunger, and inequality
- Education, skills, and research
- Climate and environment
- Finance, trade, and industry

- Health, water, and sanitation
- Governance, peace, and law
- ICT

A National Data Coordination Committee (NDCC) has been formed in 2018 comprising 50 members from ministries/divisions, public and private organizations including think tanks. This committee is chaired by the Secretary of the Statistics and Informatics Division, Government of Bangladesh. The major objectives of this committee are:

- Facilitating Bangladesh's success in the attainment of SDGs by providing required data for SDG indicators.
- Identify organizations that prepare and supply reliable, up-to-date and quality data intended for the international Indices and measurement of progress of SDGs.
- Coordinate among concerned ministries, divisions, and other organizations with regard to the statistical data computations.
- Avoid duplication in preparing the official statistics at the disaggregated level as possible.
- Assist Bangladesh Bureau of Statistics (BBS) with regard to the concept, definition, standardized procedures, and authentication of the data for the construction of international Indices.

This committee proposed the following 23 international Indices to be considered in generating required data to help better reflections in ranking:

1. Human development index	2. Global hunger index
3. Human capital index	4. ICT development index
5. Global index	6. Logistics performance index
7. E-commerce index	8. E-government development index
9. Global cyber security index	10. Human asset index
11. Gender inequality index	12. Ease of doing business index
13. Open data index	14. Network readiness index
15. Global innovation index	16. SDG index
17. happiness index	18. E-participation index
19. Global competitiveness index	20. Global connectivity index
21. Corruption perception index	22. Global peace index
23. Travel and tourism competitiveness index	

Source Bangladesh Bureau of Statistics, Dhaka, Bangladesh

The decision to undertake this initiative has important consequences in the policy making and implementation. Compared to the 169 targets of SDGs, there are a total of 527 indicators under the above 23 international Indices. On the other hand, government's election manifesto outlines 204 objectives related to the overall development of the nation. There exists overlapping among SDGs targets, international Indices, and overlapping in government's election manifesto. Hence, construction of these

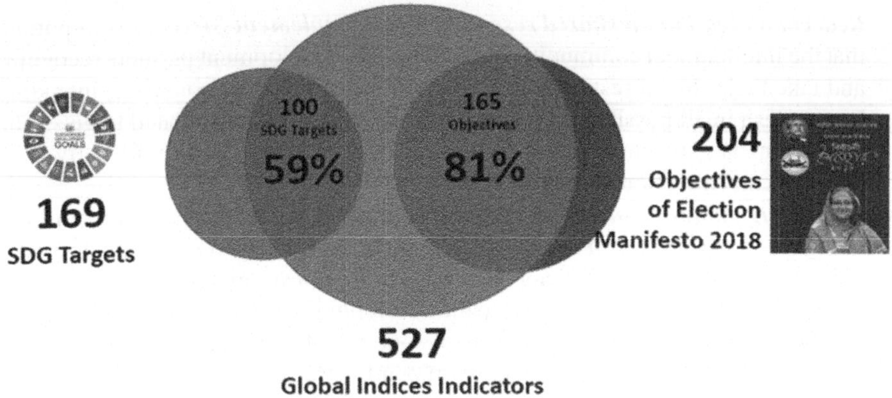

Fig. 2 Relationship of international Indices with SDGs targets and election manifesto. (*Source* Created by the author)

international Indices would help to review the progress of SDGs and objectives of the election manifesto in a timely manner. The following figure (Fig. 2) portrays the relationship of international Indices with SDGs targets and election manifesto.

6 LDC Graduation, International Cooperation, Resource Mobilization, and Capacity Development in Achieving SDG Targets

Bangladesh's LDC graduation in 2024 is likely to impact trade and international support mechanisms. In terms of trade, Bangladesh would face higher duties, especially for readymade garments (RMG). In addition, it is important to comply with the TRIPS agreement on pharmaceuticals. For development cooperation, the concessional loans available from development partners may be declined due to graduation. Due to the graduation process, Bangladesh will be parted with some LDC specific technical advantages, which include Enhanced Integrated Framework (EIF), Least Developed Countries Fund (LDCF), loss of access to LDC Technology Bank, etc. Other countries who moved from LDC status experienced slow growth, fall in ODA, but a rise in FDI. In these contexts, international cooperation is an exigency.

Forging close and dynamic partnership between Bangladesh and the development partners has never been more pressing than now, particularly in the context of implementation of SDGs and managing the upcoming graduation process. Following areas could be worthwhile for mobilizing resources and enhancing international cooperation:

- ***Redeem obligation on shared responsibility to implement SDGs***: It is important that the international community, particularly the development partners recognize and take their shared responsibility to assist the developing countries, including Bangladesh in all possible ways to achieve the SDGs and associated targets. The indications are still mixed and the development partners need to intensify their efforts in this direction on a priority basis.
- ***Mobilizing domestic and other resource***: Although the primary burden of mobilizing resources rests with the developing countries themselves through domestic means, so far the evidence is not encouraging. The development partners need to extend technical and other support to the developing countries, particularly LDCs, to enhance their capacity for domestic resource mobilization through improving tax regime, and developing other supporting infrastructures and practices to mobilize adequate domestic resources. Apart from domestic resources, ODA, FDI, and remittances could be important tools for mobilizing resources needed for development.
- ***Exploring under- or un-explored resources***: Bangladesh needs huge support for developing a sustainable and forward looking economy for the present and future generations. Exploring the Blue Economy, renewable energy, Artificial Intelligence, Automation, and biotechnology could add immense value to achieve those objectives.
- ***Opening more space for growth acceleration***: Bangladesh is on the cusp of a huge transition and would need enabling regional and global environment to grow at a steady pace during the next few decades. Support for poverty alleviation initiatives, concessional market access, assistance, FDI and loans on favorable terms, transfer of technology, and allowing space for its migrant workforce could be helpful, and the development partners, both from the developing and developed world, could contribute to this process.
- ***Extending climate management support***: Bangladesh is at the forefront of a climate disaster without contributing anything to this calamity. In the light of this, the development partners must assist Bangladesh through mitigation and adaptation support to meet those challenges on a sustainable basis.
- ***Improving capacity for management***: This is one area where perhaps the developing countries need most support from the development partners. Capacity building could cover a great deal of areas from project management to financial management to human resource development to developing legal and regulatory framework to rise productivity to competitiveness to refine dispute resolution to develop skills for conducting negotiations to advance common interests. Of course, effective capacity building and attaining lasting institutional efficiency requires attaining momentum in democratization process of governance with accountability.

Despite the immense importance of capacity building, which is deemed to be a critical element in the graduation journey of Bangladesh, there exists lack of initiatives and attention by the government to prepare its workforce to face the upcoming challenge. According to UN (2018), $33.5 billion ODA was dispensed for the purpose of capacity building and national planning in 2017, which represents 14% of total sector-allocable aid; and this share has been stable since 2010. This suggests that there are funds available for capacity building. However, the agencies need a wide array of capacity building activities and development partners or other organization have a limited mandate in terms of what they pursue with their funds. This mismatch between GoB requirements and development partner mandate is the key challenge to boost capacity building activities. To increase capacity building efforts, GoB needs to conduct an assessment on the objectives and requirements of capacity building for different implementing agencies.

7 Future Roadmaps

During the next decade, Bangladesh will have to face two challenges: (i) implementation of SDGs and targets with accelerated growth rate and (ii) management of graduation process. In addition to resources from the official sources, such as domestic tax revenue and ODA, a great deal of resources will have to be generated and mobilized from the non-governmental sources, including the private sector, CSO, and philanthropic domains. Some of the possible policy roadmaps for mobilizing resources for the implementation of SDGs and associated targets are:

- Strengthen economic diplomacy for trade growth, and for that, increasing FTAs with regional and global partners have to be expedited.
- Strengthen capacity and accountability in bureaucracy.
- Making data available for more number of indicators for both SDG targets and international Indices.
- Strengthen linkages with business community and CSOs by building effective partnerships.
- Promote public awareness and activism, etc.

Acknowledgements The author thanks an anonymous referee for making some useful suggestions in an earlier draft.

References

7th Five Year Plan FY 2016-FY 2020. (2015). *Accelerating growth, empowering citizens*. GED, Bangladesh Planning Commission.

A Handbook on Mapping of Ministries by Targets in the Implementation of SDGs aligning with 7th Five Year Plan (2016–20) (September 2016), GED, Bangladesh Planning Commission.

Data gap analysis for sustainable development goals (SDGs): Bangladesh perspective. (January 2017). GED, Bangladesh Planning Commission.

Monitoring and evaluation framework of sustainable development goals (SDGs): Bangladesh perspective. (March 2018). GED, Bangladesh Planning Commission.

National action plan of ministries/divisions by targets for the implementation of SDGs. (June 2018). GED, Bangladesh Planning Commission.
SDGs financing strategy: bangladesh perspective. (June 2017). GED, Bangladesh Planning Commission.

Some Models and Their Extensions for Longitudinal Analyses

Tapio Nummi

Abstract In this article, I present some of my statistical research in the field of longitudinal data analysis along with applications of these methods to real data sets. The aim is not to cover the whole field; rather, the perspective is based on my own personal preferences. The presented methods are mainly based on growth curve and mixture regression models and their extensions, where the focus is on continuous longitudinal data. In addition, an example of the analysis of extensive register data for categorical longitudinal data is presented. Applica-tions range from forestry and health sciences to social sciences.

1 Introduction

Longitudinal studies play an important role in many fields of science. The defining feature of these studies is that measurements of the same individual are taken repeatedly over time. The primary goal is to characterize the change in response over time as well as the factors that influence the change. Special statistical methods which address intra-individual correlation and inter-individual variation are needed. Fortunately, many statistical analysis tools developed for clustered data (e.g., mixed, multilevel, and mixture models) also apply to longitudinal data, since longitudinal data can be seen as a special case of clustered data. Here, these methods are divided into three main categories:

1. Regression and multivariate techniques,
2. Methods based on finite mixtures, and
3. Clustering techniques for categorical longitudinal data.

The main aim of this article is to briefly present some of these techniques with interesting real data applications. The purpose is not to give an overview of the topics. Instead, the perspective is based on my own research in these areas.

T. Nummi (✉)
Faculty of Information Technology and Communication Sciences, Tampere University, Tampere, Finland
e-mail: tapio.nummi@tuni.fi

B. K. Sinha et al. (eds.), *Data Science and SDGs*,
https://doi.org/10.1007/978-981-16-1919-9_2

2 Regression and Multivariate Techniques

2.1 The Growth Curve Model

Perhaps, the most important of the early models in this area is the generalized multivariate analysis of variance model (GMANOVA), which is often called the growth curve model (GCM). This model was first presented by Pothoff and Roy (1964). GCM is particularly useful in balanced experimental study designs where there are no missing data. This model can be presented as follows:

$$\mathbf{Y} = \mathbf{TBA}' + \mathbf{E},$$

where $\mathbf{Y} = (\mathbf{y}_1, \mathbf{y}_2, \ldots, \mathbf{y}_n)$ is a matrix of n response vectors, $\mathbf{T}$ is the within individual design (model) matrix, $\mathbf{B} = (\mathbf{b}_1, \ldots, \mathbf{b}_m)$ is a matrix of growth curve parameters, $\mathbf{A}$ is the between individual design matrix, and $\mathbf{E}$ is a matrix of random errors, where columns are independently normally distributed as $\mathbf{e}_i \sim N(\mathbf{0}, \boldsymbol{\Sigma})$, $i = 1, \ldots, n$.

Closed-form formulas for the estimation and testing of growth curve parameters $\mathbf{B}$ can be obtained using the Maximum Likelihood method (under unknown positive definite $\boldsymbol{\Sigma}$). Some basic results and model extensions are nicely summarized in the review papers by von Rosen (1991) and Zezula and Klein (2011).

3 Some Extensions of the Growth Curve Model

Various aspects of analysis under GCM are presented in the series of articles by myself and my co-authors. Some practical computational aspects are considered in Nummi (1989), a method for prediction is presented in Liski and Nummi (1990), an analysis under missing data with the EM algorithm is investigated in Liski and Nummi (1991), model selection for mean and covariance structure is considered in Nummi (1992), prediction and inverse estimation is investigated in Liski and Nummi (1995a), and a method of covariable selection for model parameter estimation is presented in Wang et al. (1999).

3.1 Random Effects Growth Curve Model

Perhaps, one of the most important extensions is the so-called Random effects growth curve model. This model can be written as

$$\mathbf{Y} = \mathbf{TBA}' + \mathbf{T}_c\Lambda + \mathbf{E},$$

where $\mathbf{T}_c$ is given as $\mathbf{T} = (\mathbf{T}_c, \mathbf{T}_{\bar{c}})$ and $\mathbf{\Lambda} = (\boldsymbol{\lambda}_1, \ldots, \boldsymbol{\lambda}_n)$ is a matrix of random effects. Here, we take $\mathbf{e}_i \sim N(\mathbf{0}, \sigma^2\mathbf{I})$ independent of $\boldsymbol{\lambda}_i \sim N(\mathbf{0}, \mathbf{D})$, $\forall\, i = 1, \ldots, n$, where $\mathbf{D}$ is a positive define covariance matrix.

In the article by Nummi (1997), several topics were considered: model parameter estimation and hypothesis testing, estimation under parsimonious covariance structures (e.g., AR(1)) for random errors, estimation under incomplete data using the EM algorithm, and an extension to multivariate growth curves. The multivariate extension was further studied by Nummi and Möttönen (2000), where they studied ML and REML estimation and testing in this context. For example, it was shown that under certain situations, estimated variances of growth curve parameters are greater for REML. The basic Random effects GCM was further extended and applied in small area estimation by Ngarue et al. (2017).

3.2 *Measurement Errors*

In some cases, also in an experimental situation, a measurement error may occur. This is especially the case when the planned measurement time is fixed in advance, but the exact attained measurement time does not match the planned time. An appropriate frame for this kind of analysis is found under Berkson-type measurement errors (see Berkson (1950)). The basic (Berkson) model for the observations (y, x^*) is

$$y = g(x) + \varepsilon$$

$$x = x^* + u$$

where ε and u are independent random variables with $E(\varepsilon) = E(u) = 0$. Here, the exact value of the explanatory variable x is not directly observed, but instead another quantity, the planned measurement time $x^* = x - u$ is utilized. Actually, this form of measurement error is quite common in experimental situations where the values of the predictor variable is controlled by the experimenter (e.g., in agricultural or medical studies). For GCM, the Berkson-type of measurement errors is studied in Nummi (2000) and later extended and applied to forest harvesting in Nummi and Möttönen (2004).

3.2.1 Example: Forest Harvesting

The forestry harvesting technique in the Nordic countries converts tree stems into smaller logs at the point of harvest. Modern harvesters are equipped with computer systems capable of continuously measuring the length and diameter of the stem and also predicting the profile of an unknown stem section. The harvester feeds the tree top, first through the measuring and delimbing device for a given length, then the computer predicts the rest of the stem profile and calculates the optimal cross-cutting points for the whole stem (see e.g., Uusitalo et al. (2006)). In forestry, stem curve models are often presented for relative heights (e.g., Laasasenaho 1982; Kozak 1988). However, height is unknown for a harvester, and therefore, these relative mean curve models are quite difficult to apply in practice. Low degree polynomial models were tested, e.g., in Liski and Nummi (1995b, 1996a,b) (Fig. 1).

Assume now that x is a sum of sub-intervals $\delta_i^{\ *} = \delta_i + \xi_i$

$$
\begin{aligned}
x &= \sum \delta_i^* \\
&= \sum \delta_i + \sum \xi_i \\
&= x^* + u
\end{aligned}
$$

where random error u is a sum of random terms $u = \sum \xi_i$, where ξ_i are independent with $E(\xi_i) = 0$ and $\mathrm{Var}(\xi_i) = \sigma^2_{\xi_i}$. Random errors u are now dependent and the variance σ_u^2 increases with x^*. A special model for the covariance structure $\mathrm{Var}(\mathbf{y})$ is now needed.

The general least squares methods provide unbiased parameter estimates only in the most simple first-degree model $g(x^*) = \beta_0 + \beta_1 x^*$. For more complex models, for $g(x^*)$, the least squares estimates of the model parameters are biased. However,

Fig. 1 A forest harvester at work. In the figure, the harvester has cut down a tree and started pruning the branches. At the same time, the stem diameter and length are measured and the measurements are transferred to the harvester's computer. The harvester is now at the first possible cutting point, at which the decision must be made as to whether to cut at that point or at another point along the stem. *Source* Ponsse company, Finland

as shown in Nummi and Möttönen (2004), predictions may still be unbiased for low-degree polynomial models. Estimation and prediction for an extended Berkson model (with dependent measurement errors) are considered in Nummi and Möttönen (2004).

3.3 *Spline Growth Model*

A more general formulation of the basic GMANOVA can be written as

$$\mathbf{Y} = \mathbf{GA}' + \mathbf{E}$$

where $\mathbf{G} = (\mathbf{g}_1, \ldots, \mathbf{g}_m)$ is a matrix of smooth mean curves (Spline Growth Model, SGM; Nummi and Koskela (2008); Nummi and Mesue 2013; Mesue and Nummi 2013; Nummi et al. 2017). Here we assume that

$$\Sigma = \sigma^2 \mathbf{R}(\boldsymbol{\theta})$$

where $\mathbf{R}$ takes a certain kind of parsimonious covariance structure with covariance parameters $\boldsymbol{\theta}$. A smooth solution for $\mathbf{G}$ can be obtained by minimizing the penalized least squares criterion (see Nummi and Koskela 2008)

$$\text{PLS} = \text{tr}\big[\big(\mathbf{Y} - \dot{\mathbf{G}}\big)\mathbf{H}\big(\mathbf{Y} - \dot{\mathbf{G}}\big) + \alpha\dot{\mathbf{G}}'\mathbf{K}\dot{\mathbf{G}}\big],$$

where α (>0) is a fixed smoothing parameter, $\dot{\mathbf{G}} = \mathbf{GA}'$, $\mathbf{H} = \mathbf{R}^{-1}$, and the roughness matrix $\mathbf{K}$ (from $RP = \int g\prime\prime^2$) is defined as $\mathbf{K} = \nabla\Delta^{-1}\nabla'$ where ∇ and Δ are banded $q \times (q-2)$ and $(q-2) \times (q-2)$ matrices defined as (non-zero elements)

$$\nabla_{l,l} = \frac{1}{h_l},\ \nabla_{l+1,l} = -\left(\frac{1}{h_l} + \frac{1}{h_{l+1}}\right),\ \nabla_{l+2,l} = \frac{1}{h_{l+1}}$$

and

$$\nabla_{l,l+1} = \nabla_{l+1,l} = \frac{l_{k+1}}{6},\ \nabla_{l,l} = \frac{h_l + h_{l+1}}{3},$$

where $h_j = t_{j+1} - t_j$, $j = 1, 2, \ldots, (q\text{-}1)$ and $l = 1, 2, \ldots, (q - 2)$ (see e.g., Green and Silverman (1994)).

As shown in Nummi and Koskela (2008), the minimizer is easily seen by rewriting the PLS-function in a slightly different form. Then given α and $\mathbf{H}$, the spline estimator becomes

$$\tilde{\mathbf{G}} = (\mathbf{H} + \alpha\mathbf{K})^{-1}\mathbf{HYA}(\mathbf{A}'\mathbf{A})^{-1},$$

where the fitted growth curves $\tilde{\mathbf{G}}$ are natural cubic smoothing splines. It is further easily seen that if $\mathbf{K} = \mathbf{KH}$ (or $\mathbf{KR} = \mathbf{K}$), the spline estimator simplifies as

$$\hat{\mathbf{G}} = \mathbf{SYA}(\mathbf{A}'\mathbf{A})^{-1},$$

where the so-called smoother matrix is denoted as $\mathbf{S} = (\mathbf{I} + \alpha\mathbf{K})^{-1}$. Note that this is an important simplification, since estimates $\hat{\mathbf{G}}$ are now simple linear functions of the observations (α fixed). In a sense, this can be compared to the results of linear models, where OLSE = BLUE. It is quite easy to see that certain important structures (e.g., uniform, random effects, etc.) used for the analysis of longitudinal data meet the simplifying condition. The smoothing parameter α can then be chosen using the Generalized Cross-Validation (GCV) criteria, for example.

3.3.1 Testing with an Application for Behavioral Cardiology

Note that the smoother matrix $\mathbf{S}$ is not a projection matrix, and therefore, certain results developed for linear models are not directly applicable for SGM. Our approach is to approximate $\mathbf{S}$ with the following decomposition (Nummi and Mesue 2013; Mesue and Nummi 2013; Nummi et al. 2017 and Nummi et al. 2018)

$$\mathbf{S} = \mathbf{M}(\mathbf{I} + \alpha\Lambda^{-1})\mathbf{M}',$$

where $\mathbf{M}$ is the matrix of q orthogonal eigenvectors of $\mathbf{K}$ and $\mathbf{\Lambda}$ is a diagonal matrix of corresponding eigenvalues obtained from the Spectral decomposition. Here, we assume that eigenvectors are ordered according to eigenvalues $\gamma = 1/(1 + \alpha\lambda)$ of $\mathbf{S}$, where λ is an eigenvalue of $\mathbf{K}$. Note that the sequence of eigenvectors $\mathbf{m}_1$, $\mathbf{m}_2$, ..., $\mathbf{m}_q$ increases in complexity. The first two eigenvectors, $\mathbf{m}_1$, $\mathbf{m}_2$, span a straight line model and the corresponding eigenvalues are 1.

Using the approximation $\mathbf{S} \approx \mathbf{M}_c\mathbf{M}_c'$ the set of fitted curves with SGM are

$$\bar{\mathbf{Y}} = \mathbf{M}_c\mathbf{M}_c'\mathbf{Y}\mathbf{A}\left(\mathbf{A}'\mathbf{A}\right)^{-1}\mathbf{A}' = \mathbf{M}_c\hat{\Omega}\mathbf{A}',$$

where $\mathbf{M}_c$ is a matrix of c first eigenvectors of $\mathbf{S}$ that can be chosen using GCV criteria, and $\hat{\Omega} = \mathbf{M}_c'\mathbf{Y}\mathbf{A}\left(\mathbf{A}'\mathbf{A}\right)^{-1}$. All the relevant information for testing is now in $\hat{\Omega}$, which can be seen to be unbiased estimates of the parameters of the growth curve model

$$E(\mathbf{Y}) = \mathbf{M}_c\Omega\mathbf{A}'.$$

Testing can be based on the linear hypothesis of the form H_0: $\mathbf{C\Omega D} = \mathbf{O}$, where $\mathbf{C}$ and $\mathbf{D}$ are appropriate given matrices. It is easy to construct an F-test for this H_0. It is further easy to show that for some important special cases, the distribution of this F-test does not depend on the estimated covariance structure.

For example, in Tampere Ambulatory Hypertension Study, 95 men (aged 35–45 years with the same ethnic and cultural background) were selected and their body functions were accurately monitored for one day (see Nummi et al. 2017). Inclusion criteria—healthy according to conventional health criteria and not on medication. For this study, we investigated the hourly means of systolic blood pressure (SBP), diastolic blood pressure (DBP), and heart rate (HR). The participants were classified before the experiment into two groups:

Group 1: Normotensive (NT), 33 participants, and.

Group 2: Borderline hypertensive (BHT) and hypertensive (HT), 62 participants.

In our example, we were especially interested in testing whether blood pressure variables behaved the same during the day in these two groups. In particular, by definition, there is a level difference in these two groups. If we define the roughness matrix as $\mathbf{K}_s = \mathbf{W} \otimes \mathbf{K}$, where $\mathbf{W}$ = diag $(\alpha_1, \ldots, \alpha_s)$ the multivariate uniform structure

$$\mathbf{R} = \left(\mathbf{I}_s \otimes \mathbf{1}_q\right)\mathbf{D}\left(\mathbf{I}_s \otimes \mathbf{1}_q\right)' + \mathbf{I}_{qs}$$

where $\mathbf{D}$ is a covariance matrix, satisfies the multivariate version of the simplifying condition $\mathbf{RK}_s = \mathbf{K}_s$ and the unweighted spline estimator becomes (Mesue and Nummi 2013)

$$\hat{\mathbf{G}} = (\mathbf{I}_{qs} + \mathbf{W} \otimes \mathbf{K})^{-1} \mathbf{YA}(\mathbf{A}'\mathbf{A})^{-1}$$

$$= \begin{pmatrix} \mathbf{S}(\alpha_1) & 0 & 0 \cdots & 0 \\ 0 & \mathbf{S}(\alpha_2) & 0 \cdots & 0 \\ \vdots & \vdots & \vdots \ddots & \vdots \\ 0 & 0 & 0 \cdots & \mathbf{S}(\alpha_s) \end{pmatrix} \mathbf{YA}(\mathbf{A}'\mathbf{A})^{-1} \tag{1}$$

where $\mathbf{S}(\alpha_j) = (\mathbf{I}_q + \alpha_j \mathbf{K})^{-1}$, where α_j is a smoothing constant for $j = 1, \ldots, s$. A straightforward generalization of the earlier considerations gives us an estimator

$$\hat{\Omega} = \mathbf{M}'_{\bullet} \mathbf{YA}(\mathbf{A}'\mathbf{A})^{-1},$$

where $\boldsymbol{M}_{\bullet} = \mathrm{diag}(\mathbf{M}_1, \ \mathbf{M}_2, \ \ldots, \ \mathbf{M}_s)$ and the corresponding multivariate growth curve model is

$$\mathbf{Y} = \mathbf{M}_{\bullet} \Omega \mathbf{A}'.$$

Testing (see Nummi et al. 2017) can be based on the linear hypothesis H_0: $\mathbf{C\Omega D} = \mathbf{0}$, where $\mathbf{C}$ and $\mathbf{D}$ are known $v \times c$ and$m \times g$ matrices with ranks v and g, respectively, with

$$F = \frac{Q_*/vg}{\hat{\sigma}^2} \sim F\,[vg, \ n(sq - c_{\mathrm{tot}})],$$

where $c_{\mathrm{tot}} = c_1 + \ldots + c_s$ and

$$Q_* = \mathrm{tr}\left[\mathbf{D}'(\mathbf{A}'\mathbf{A})^{-1}\mathbf{D}\right]^{-1}\left[\mathbf{C}\hat{\Omega}\mathbf{D}\right]'\left[\mathbf{C}\mathbf{M}'_{\bullet}\mathbf{R}\mathbf{M}_{\bullet}\mathbf{C}'\right]^{-1}\left[\mathbf{C}\hat{\Omega}\mathbf{D}\right]$$

and

$$\hat{\sigma}^2 = \sum_{l=1}^{s} \frac{1}{n(q - c_l)} \mathrm{tr}\mathbf{Y}'_l(\mathbf{I}_q - \mathbf{P}_l)\mathbf{Y}_l \, .$$

Often $\mathbf{R}$ may not be known and need to be estimated. In this case, the distribution of F is only approximate. However, with the multivariate uniform covariance model, when investigating the progression only we can take $\mathbf{C} = [\mathbf{I}_s \otimes (\mathbf{0}, \mathbf{I})]$. It can then be shown that the test statistics have an exact F-distribution.

For the example data, $c_1 = 12$ (SBP), $c_2 = 10$ (DPB), and $c_3 = 12$ (HR). To test if the progression is the same in both groups, we attained

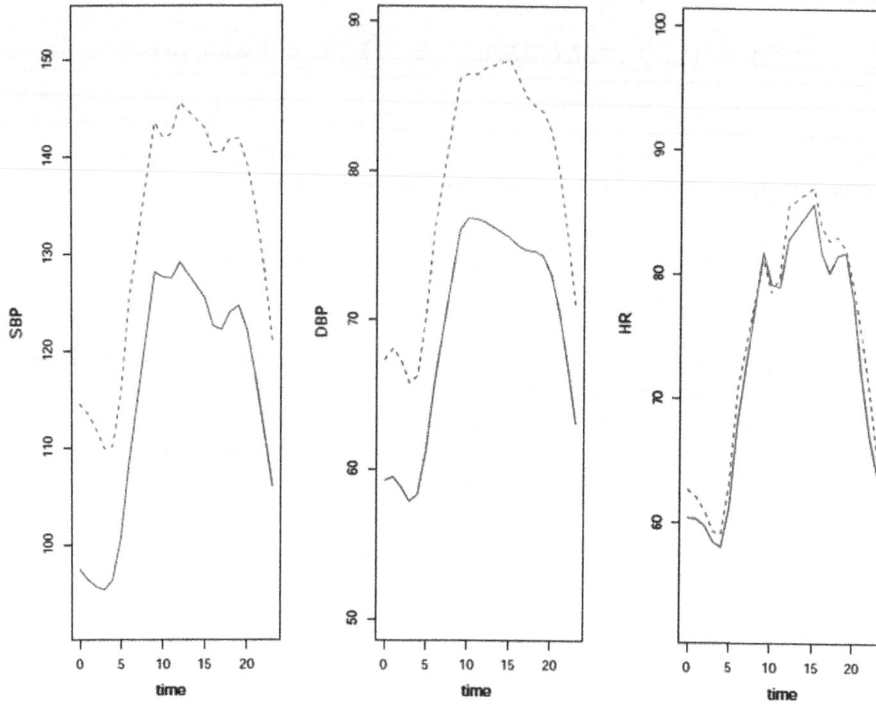

Fig. 2 Fitted mean curves for systolic blood pressure (SBP), diastolic blood pressure (DBP), and heart rate (HR) during the test day. Solid line is for Group 1 (normotensive) and dotted line for Group 2 (Borderline hypertensive and hypertensive). *Source* Created by the authors

$$F = \frac{1811.041/31}{66.26201} = 0.88166,$$

which gives the P-value $P\left(F_{31,2470} \geq 0.88166\right) \approx 0.654967$. Therefore, the null hypothesis of equal progression for each of the variables in these groups is not rejected. The fitted mean curves are shown in Fig. 2.

4 Models Based on Finite Mixtures

4.1 Introduction

Denote random vectors of longitudinal measurements as $\mathbf{y}_i = \left(y_{i1}, \ y_{i2}, \ \ldots, \ y_{ip_i}\right)'$ and the marginal probability density of $\mathbf{y}_i$ with possible time-dependent covariates $\mathbf{X}_i$ as $f(\mathbf{y}_i|\mathbf{X}_i)$ for $i = 1, \ldots, N$. It is assumed that $f\ (\mathbf{y}_i|\mathbf{X}_i)$ follows a mixture of K densities

$$f(\mathbf{y}_i|\mathbf{X}_i) = \sum_{k=1}^{K} \pi_k f_k(\mathbf{y}_i|\mathbf{X}_i), \qquad \sum_{k=1}^{K} \pi_k = 1 \text{ with } \pi_k > 0,$$

where π_k is the probability of belonging to the cluster k and $f_k(\mathbf{y}_i\,|\mathbf{X}_i)$ is the density for the kth cluster. If the multivariate normal distribution is assumed, we have

$$f_k(\mathbf{y}_i\,|\mathbf{X}_i) = (2\pi)^{\frac{1}{2}}|\Sigma_{ik}|^{\frac{p_i}{2}} \exp\left\{\frac{1}{2}(\mathbf{y}_i - \boldsymbol{\mu}_{ik})'\Sigma_{ik}^{-1}(\mathbf{y}_i - \boldsymbol{\mu}_{ik})\right\},$$

where $\boldsymbol{\mu}_{ik}(\boldsymbol{\theta}_k, \mathbf{X}_i)$ is a function of covariates $\mathbf{X}_i$ with parameters $\boldsymbol{\theta}_k$, and $\boldsymbol{\Sigma}_{ik}(\sigma_k)$ is a variance–covariance matrix within the kth cluster, involving a vector of unique covariance parameters $\boldsymbol{\sigma}_k$. In the most general case, $\boldsymbol{\mu}_k$ and $\boldsymbol{\Sigma}_k$ are the unstructured mean and covariance matrices. However, often some more parsimonious structures are imposed either on $\boldsymbol{\mu}_k$ or $\boldsymbol{\Sigma}_k$, or on both.

In this article, we focus on the normal GLM case (so-called trajectory analysis; see e.g., Nagin 1999, 2005). It is then simply assumed that

$$\boldsymbol{\mu}_k = \mathbf{X}_i\boldsymbol{\theta}_k \quad \text{with} \quad \Sigma_k = \sigma_k^2\mathbf{I}\,.$$

Note that this conditional independence assumption does not mean independence over the whole sample. One important aim is to identify and estimate the possible subpopulations as well as possible. Therefore, often a natural interpretation of the identified groups is emphasized jointly with model selection and statistical fit criteria.

4.2 *Data Analysis: Analysis of Drinking Profiles*

The Northern Swedish Cohort Study covers all pupils who in 1981 attended the final year of compulsory school (at age 16) in Luleå. The number of participants in all follow-up surveys (1983, 1986, 1995, and 2007) was 1,005. For this study, we used the alcohol consumption (converted to absolute alcohol in centiliters) of male participants at the age of 16, 18, 21, 35, and 42 years (for more details, see Virtanen et al. (2015)).

The distribution of alcohol consumption is highly skewed, but there is a relatively high probability of zero observations (semicontinuous data). This kind of data is quite common in many areas (cf. zero inflation). One possible solution to the problem is to apply mixture modeling, where one component of the mixture functions is degenerated near to zero. The advantage is that the mixture approach allows additional heterogeneity by avoiding a sharp dichotomy between zero and near-zero observations. A brief summary of the methods for semicontinuous and zero-inflated data is presented in Min and Agresti (2005).

Table 1 Results of the fits of the mixture models with $k = 1, 2, \ldots, 7$ and $\lambda \in [-2, 2]$

K	$\hat{\lambda}$	L	BIC	AIC
1	0.16	−16,367.23	32,770.40	32,744.47
2	0.04	−16,065.51	32,206.09	32,151.02
3	0.00	−15,961.67	32,037.55	31,953.35
4	0.08	−15,885.47	31,924.28	31,810.94
5	0.08	−15,852.29	31,897.07	31,754.58
6	0.04	−15,818.75	31,869.12	31,697.49
7	0.04	−15,802.01	31,874.97	31,674.20

Source Authors' own processing

Here the so-called "broken stick" model was applied as the basic model for transformed observations:

$$Y^{(\lambda)} = \beta_0 + \beta_1 t + \beta_2 (t - k_1)_+ + \varepsilon,$$

$k_1 = 21$, $(.)_+$ equals (.) if $(.) \geq 0$ and 0 otherwise. The model consists of two combined straight lines at the age of 21.

The number of clusters K and the transformation parameter of the Box-Cox transformation was jointly estimated. Depending on the criterion, a different K is selected. With AIC, the minimum is obtained for $K = 7$, and with BIC, the minimum is obtained for $K = 6$ (Table 1). Our choice was $K = 6$, which was also in line with the earlier studies with these data (e.g., Virtanen et al. (2015)). The estimate of the transformation parameter (Box-Cox) $\hat{\lambda} = 0.04$ suggested the log transformation.

The estimated mixture proportions are $\pi_1 = 0.1558$; $\pi_2 = 0.1034$; $\pi_3 = 0.0737$; $\pi_4 = 0.2677$; $\pi_5 = 0.2526$; and $\pi_7 = 0.1468$. Group 3 is the zero or near-zero cluster. Interestingly, those who had a high consumption level at the earlier ages tended to maintain a high consumption level also at the later ages (Fig. 3).

4.3 *Extension: Semiparametric Mean Model*

The set of explanatory variables in $\mathbf{X}_i$ is divided into the parametric part $\mathbf{U}_i$ and the non-parametric part $\mathbf{t}_i$, where $\mathbf{t}_i$ is the vector of measuring times $t_1, \ldots, t_{p_i}$. For the ith individual within the kth cluster, we assume the semiparametric model

$$\mathbf{y}_{ik} = \mathbf{g}_{ik}(\mathbf{t}_i) + \mathbf{U}_i \mathbf{b}_k + \epsilon_i,$$

where $g_{ik}(\mathbf{t}_i)$ is a smooth vector of twice differentiable functions evaluated at time points $\mathbf{t}_i$, $\mathbf{U}_i$ is a matrix of h covariates (constant term not included), $\mathbf{b}_k$ is a parameter vector to be estimated, and $\mathrm{Var}(\epsilon_i) = \sigma_k^2 \mathbf{I}_i$.

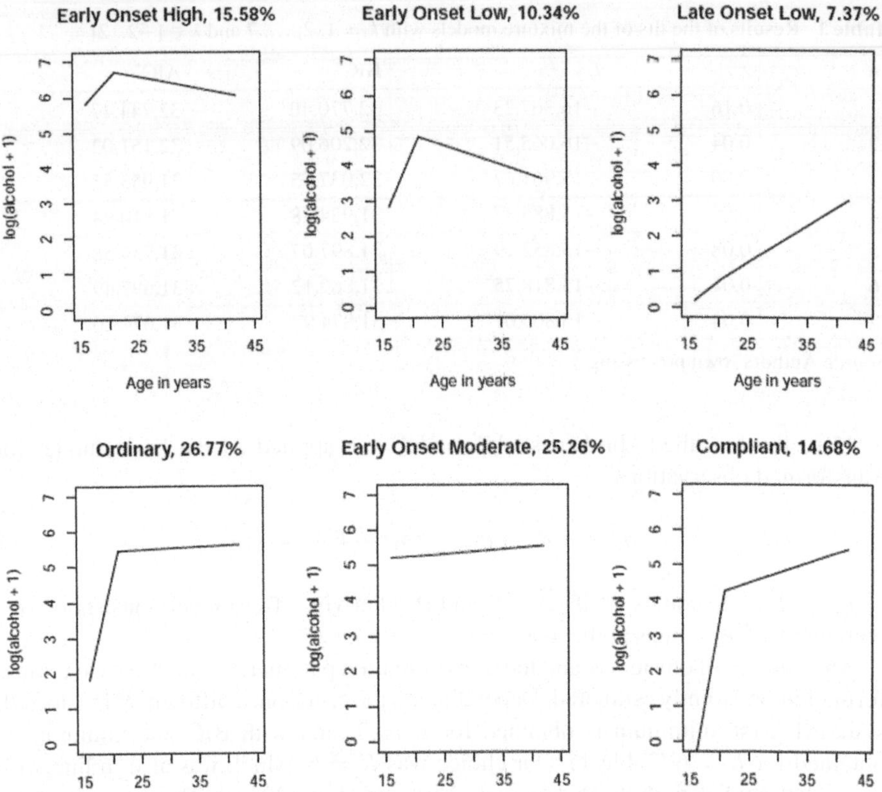

Fig. 3 Fitted trajectory curves for alcohol consumption of males. *Source* Created by the authors

When using the EM algorithm, the estimation problem can be seen as a missing data problem, where $\mathbf{y}_i$ are observed but "group indicators" $\mathbf{z}_i'$ are missing. We denote

$$\mathbf{y}_i^* = (\mathbf{y}_i', \mathbf{z}_i')'$$

where $z_{ik} = 1$ if $\mathbf{y}_i$ stemmed from the kth component; otherwise, $z_{ik} = 0$. The vectors $\mathbf{z}_1,..., \mathbf{z}_N$ can now be seen as realized values of random vectors $\mathbf{Z}_1,..., \mathbf{Z}_N$ from the multinomial distribution. The complete-data, joint penalized log-likelihood function is (see Nummi et al. (2018) for details)

$$l_c(\phi) = \sum_{i=1}^{N} \sum_{k=1}^{K} \left\{ z_{ik} \left[\log(\pi_k) + \log(f_k) \right] - \frac{\alpha_k}{2N} \mathbf{g}_k' \mathbf{K} \mathbf{g}_k \right\}.$$

The E step is to calculate

$$E\left(Z_{ik}|\hat{\phi},\ \mathbf{y}_1,\ \ldots,\ \mathbf{y}_N\right) = \frac{\hat{\pi}_k f_k\left(\mathbf{y}_i|\mathbf{X}_i,\ \hat{\boldsymbol{\xi}}_k\right)}{\sum_{l=1}^{k} \hat{\pi}_l f_l\left(\mathbf{y}_i|\mathbf{X}_i,\ \hat{\boldsymbol{\xi}}_l\right)} = \hat{z}_{ik}$$

where $\hat{\boldsymbol{\xi}}_1, \ldots, \hat{\boldsymbol{\xi}}_K$ are vectors consisting of estimates of mixing distribution mean and variances. In the M step, the expected log-likelihood for the completed data

$$E[l_c(\phi)] = \sum_{i=1}^{N}\sum_{k=1}^{K}\left\{\hat{z}_{ik}\left[\log(\pi_k) + \log(f_k)\right] - \frac{\alpha_k}{2N}\mathbf{g}_k'\mathbf{K}\mathbf{g}_k\right\}$$

is maximized. These two steps are iterated until convergence. The method gives closed-form formulas for $\mathbf{g}_k$ and $\mathbf{b}_k$ with estimates for π_k. Here each of the k group can be smoothed independently, and thus this provides a very flexible model within each of the k clusters. In Nummi et al. (2018), a technique providing an approximate solution is also introduced. This makes semiparametric mixture analysis possible in general statistical software developed for mixture regression.

5 Clustering Techniques for Categorical Longitudinal Data: Factory Downsizing

Example of sequence analysis is based on Statistics Finland's combined employee–employer data (FLEED), which includes data for all 15–70 year old of those who lived in Finland in 1988–2014. For research purposes, a random sample of the size of one-third was taken. The starting point of the study group is those enterprises that reduced more than 30% of staff or were dismissed in the year 2005. The actual study group taken then consisted a sample of 7,730 people (aged 45–60) who lost their job in 2005 (followed until 2014). A reference group of matched (Propensity score) 7,844 people from the same register who did not lose their job in 2005 was also taken. Since the data are categorical (employment status), so-called sequence analysis was applied to the combined data. Sequence analysis was performed with R software using the Weighted Cluster library Studer (2013). The number of clusters was evaluated using the Average Silhouette Width. For further details on the data, methods, and results, we refer to Kurvinen et al. (2018).

Finally, six clusters were identified that were named according to the main activity prevailing in the group. The results are presented in Table 2. It is observed that in the study group about half of the sample still continued in employment. It can be seen as an indication of an effective labor market policy in Finland. However, there is clearly an elevated risk (compared to those who continued as employed) for those who lost their job entering the unemployment group and the unemployment pension group even after controlling for covariates gender, age, sector of employment, education,

Table 2 Results of the sequence analysis of combined employee–employer data

Main activity	Combined (%)	Study group (%)	Reference (%)
Employed	52	49	55
Disability pension	12	13	11
Retired	12	10	13
Part-time retired	10	9	11
Unemployed	9	11	7
Unemp. pension	5	8	3

Source Authors' own processing

socio-economic status, type of residence area, employment and unemployment in 2004, and sickness allowance paid in 2003–2004.

Although sequence analysis is mainly descriptive in nature, a suitable experimental study design can also provide a framework for further estimation and testing of important statistical quantities.

Acknowledgements The author thanks an anonymous referee for making some useful comments and suggestions in an earlier draft.

References

Berkson, J. (1950). Are there two regressions? *Journal of the American Statistical Association, 45*, 164–180.

Green, P., & Silverman, B. (1994). *Nonparametric regression and generalized linear models. A roughness penalty approach* (58th ed.). Monographs on Statistics and Applied Probability, Boca Raton, FL: Chapman Hall/CRC.

Kozak, A. (1988). A variable-exponent taper equation. *Canadian Journal of Forest Research, 18*, 1363–1368.

Kurvinen, A., Jolkkonen, A., Koistinen, P., Lipiäinen, L., Nummi, T. & Virtanen, P. (2018). Työn menetys työuran loppuvaiheessa—Tutkimus 45–60-vuotiaana rakennemuutoksessa työnsä menettäneiden työllisyysurista ja riskistä päätyä työttömäksi tai työvoiman ulkopuolelle, 83, 5–6. (in Finnish with English abstract).

Laasasenaho, J. (1982). Taper curve and volume functions for pine, spruce and birch. *Communicationes Instituti Forestalis Fenniae, 108*. http://urn.fi/URN:ISBN:951-40-0589-9

Liski, E. P., & Nummi, T. (1990). Prediction in growth curve models using the EM algorithm. *Computational Statistics & Data Analysis, 10*(2), 99–108.

Liski, E. P., & Nummi, T. (1995a). Prediction and iinverse estimation in repeated-measures models. *Journal of Statistical Planning and Inference, 47*, 141–151.

Liski, E. P., & Nummi, T. (1995b). Prediction of tree stems to improve efficiency in automatized harvesting of forests. *Scandinavian Journal of Statistics, 22*(2), 255–269.

Liski, E. P., & Nummi, T. (1996a). Prediction in repeated-measures models with engineering applications. *Technometrics, 38*(1), 25–36. https://doi.org/10.1080/00401706.1996.10484413

Liski, E. P., & Nummi, T. (1996b). The marking for bucking under uncertainty in automatic harvesting of forest. *International Journal of Production Economics, 46–47*, 373–385.
Liski E. P., & Nummi T. (1991). Missing data under the GMANOVA model. In Özturk, & van der Meulen (Eds.), *Conference Proceedings on the Frontiers of Statistical Scientific Theory—Industrial Applications (Volume II of the Proceedings of ICOSCO-I, The First International Conference on Statistical Computing)* (pp. 391–404). Columbus, Ohio: American Science Press Inc.
Mesue, N., & Nummi, T. (2013). Testing of growth curves using smoothing spline: A multivariate approach. In V. M. R. Muggeo, V. Capusi, G. Boscaino, & G. Lovison (Eds.), *Proceedings of the 28th International Workshop on Statistical Modelling* (pp. 281–288). Palermo, Italia: Statistical Modelling Society.
Min, Y., & Agresti, A. (2005) Random effect models for repeated measures of zero-inflated count data. *Statistical Modelling, 5*(1), 1–19.
Nagin, D. S. (1999). Analyzing developmental trajectories: A semiparametric, group-based approach. *Psychological Methods, 4*(2), 139–157. https://doi.org/10.1037/1082-989X.4.2.139
Nagin, D. S. (2005). *Group-based modeling of development*. Harvard University Press.
Ngaruye, I., Nzabanita, J., von Rosen, D., & Singull, M. (2017). Small area estimation under a multivariate linear model for repeated measures data. *Communications in Statistics—Theory and Methods, 46*(21), 10835–10850. https://doi.org/10.1080/03610926.2016.1248784
Nummi, T. (1997). Estimation in a random effects growth curve model. *Journal of Applied Statistics, 24*(2), 157–168. https://doi.org/10.1080/02664769723774
Nummi, T. (2000). Analysis of growth curves under measurement errors. *Journal of Applied Statistics, 27*(2), 235–243. https://doi.org/10.1080/02664760021763
Nummi, T., & Koskela, L. (2008). Analysis of growth curve data using cubic smoothing splines. *Journal of Applied Statistics, 35*(6), 681–691. https://doi.org/10.1080/02664760801923964
Nummi, T., & Möttönen, J. (2000). On the analysis of multivariate growth curves. *Metrika, 52*, 77–89. https://doi.org/10.1007/s001840000063
Nummi, T. (1989). APL as a tool for computations in growth studies. In A. Kertesz, L. C. Shaw (Eds.), *APL Quote-Quad (Conference Proceedings APL89 - APL as a Tool of Thought, August 7–10, New York City), Volume 19, Number 4* (pp. 293–298).
Nummi, T. (1992). On model selection under the GMANOVA model. In P.G. M van der Heyden, W. Jansen, B. Francis, & G. U. H. Seeber (Eds.), *Statistical Modelling* (pp. 283–292). Amsterdam: Elsevier Science Publishers B.V.
Nummi, T., & Mesue, N. (2013). Testing of growth curves with cubic smoothing splines. In R. Dasgupta (Ed.), *Advances in Growth Curve Models. Springer Proceedings in Mathematics & Statistics, vol. 46* (pp. 49–59). Springer, New York. https://doi.org/10.1007/978-1-4614-6862-2_3
Nummi, T., & Möttönen, J. (2004). Estimation and prediction for low-degree polynomial models under measurement errors with an application to forest harvester. *Journal of the Royal Statistical Society: Series C Applied Statistics, 53*, 495–505. https://doi.org/10.1111/j.1467-9876.2004.05138.x
Nummi, T., Möttönen, J., & Tuomisto, M. T. (2017). Testing of multivariate spline growth model. In D. G. Chen, Z. Jin, G. Li, Y. Li, A. Liu, Y. Zhao (Eds.), *New Advances in Statistics and Data Science* (pp. 75–85). Springer, Cham: ICSA Book Series in Statistics. https://doi.org/10.1007/978-3-319-69416-0_5.
Nummi T., Salonen J., Koskinen, L., & Pan, J. (2018). A semiparametric mixture regression model for longitudinal data. *Journal of Statistical Theory and Practice, 12*(1), 12–22. https://doi.org/10.1080/15598608.2017.1298062.
Potthoff, R. F., & Roy, S. N. (1964). A generalized multivariate analysis of variance model useful especially for growth curve problems. *Biometrika, 51*(3–4), 313–326. https://doi.org/10.1093/biomet/51.3-4.313
Studer, M. (2013) *Weighted cluster library manual: Apractical guide to creating typologies of trajectories in the social sciences with R. LIVES* (Working Paper No. 24). https://doi.org/10.12682/lives.2296-1658.2013.24

Uusitalo, J., Puustelli, A., Kivinen, V. -P., Nummi, T., & Sinha, B. K. (2006). Bayesian estimation of diameter distribution during harvesting. *Silva Fennica, 40*(4), 663–671.

Virtanen, P., Nummi, T., Lintonen, T., Westerlund, H., Hägglöf, B., & Hammarström, A. (2015). Mental health in adolescence as determinant of alcohol consumption trajectories in the Northern Swedish cohort. *International Journal of Public Health, 60*, 335–342. https://doi.org/10.1007/s00038-015-0651-5

von Rosen, D. (1991). The growth curve model: A review. *Communications in Statistics—Theory and Methods, 20*(9), 2791–2822. https://doi.org/10.1080/03610929108830668

Wang, S. G., Liski, E. P., & Nummi, T. (1999). Two-way selection of covariables in multivariate growth curve models. *Linear Algebra and Its Applications, 289*, 333–342.

Zezula, I., & Klein, D. (2011). Overview of recent results in growth-curve-type multivariate linear models. *Mathematica, 50*(2), 137–146.

Association of IL-6 Gene rs1800796 Polymorphism with Cancer Risk: A Meta-Analysis

Md. Harun-Or-Roshid, Md. Borqat Ali, Jesmin, and Md. Nurul Haque Mollah

Abstract Interleukin-6 (IL-6) gene polymorphisms are a crucial functional marker in human body. Several genetic association studies reported the significant association between IL-6 gene and various major disease and cancers. In this study, the association of IL-6 gene polymorphism (rs1800796) with cancer risk was investigated through a meta-analysis, which included the larger sample size. To find the association between IL-6 gene (−572 G/C) polymorphism, we extracted the dataset in 27 eligible studies for 24,138 subjects through an efficient searching strategy from PubMed, PubMed central, web of science, google scholar, and other relevant biological literature-based online databases until February 2019. We investigated the association by comparing the allelic and genotypic case–control frequency based on odds ratio with 95% confidence interval and some other statistical tests. According to the results, the rs1800796 SNP significantly associated with increasing risk (CG vs. CC + GG: OR = 1.12, 95% CI = 1.01 – 1.23, p = 0.0288) of overall cancer, particularly with lung, stomach, and prostate cancer as well as for Asian ethnicity. These findings suggest that IL-6 gene polymorphisms may appraise as a genetic biomarker for cancer risks.

Keywords Association · Interleukin-6 gene · Cancer risk · Case–Control studies · Sample size · Meta-analysis

---

Md. Harun-Or-Roshid · Md. B. Ali · Md. N. H. Mollah (✉)
Department of Statistics, University of Rajshahi, Rajshahi, Bangladesh
e-mail: mollah.stat.bio@ru.ac.bd

Jesmin
Department of Genetic Engineering and Biotechnology, University of Dhaka, Dhaka, Bangladesh
e-mail: jesmin@univdhaka.edu

B. K. Sinha et al. (eds.), *Data Science and SDGs*,
https://doi.org/10.1007/978-981-16-1919-9_3

1 Introduction

According to World Health Organization (WHO), the global cancer burden is estimated to have risen to 18.1 million new cases and 9.6 million deaths in 2018 (WHO 2018). According to the GLOBOCAN database in 2018, they estimate the extent of mortality and outbreak from 36 types of cancer and for all cancer in 185 countries in the world (IACR, http://gco.iarc.fr/). Cancer is a life-threatening disease in human life due to its abnormal cell growth and spreading anywhere in the body. So it is very important for us to stand against cancer by serviceable research.

Interleukin-6 (IL-6) is a group of protein that acts as pro-inflammatory and anti-inflammatory of cytokine and myokine, respectively. IL-6 have foreword in tumor microenvironment regulation (Li et al. 2012), production of stem cell-like cells for increasing the risk of breast cancer (Xie et al. 2012), transference to down-regulate the tumor suppressor protein (E-cadherin) (Miao et al. 2014), and obstruction for diagnosis in oral cancer (Gasche et al. 2011). The IL-6 was considered in heavy cancer patients with intensive level (Medscape.com 2006) and showing the strong association of cancer with inflammation of tumor microenvironment (Kumari et al. 2016). Moreover, the circulation of IL-6 was associated with short overall survival (OS) in gastrointestinal cancer (GI) patients (Vainer et al. 2018). For preventing the severity of cancer, it might be needed the blocking of IL-6 or inhibiting its associated signal or applying the combination of traditional conventional anti-cancer therapies (Kumari et al. 2016).

The previous meta-analysis study performed by Zhou et al. (2014) reported no significant association of IL-6 gene with the risk of colon cancer. Another meta-analysis by Wang et al. (2014) revealed the insignificant association of -572G/C polymorphism with lung cancer, and Zhou et al. (2016) also have shown in same line for the gene polymorphisms. The recent meta-analysis by Zhou et al. (2018) analyzed three polymorphisms and proposed that the significant association of rs1800796 polymorphism with overall cancer risk only. Therefore, we also conducted the study for this polymorphisms of IL-6 gene with a greater sample size from existing literature, and investigated the association by overall and subgroup analysis through each type of cancer. Besides the exploration, we also notified the inference of allele and detailed the results in polymorphism.

2 Methods and Materials

2.1 Search Strategy

We searched the pertinent articles and pulled together the literature from PubMed, PubMed Central, Google Scholar, Web of Science, and other online databases published until February 2019. For searching the articles, we used the following keywords: (i) IL-6, (ii) IL-6, Cancer, (iii) IL-6, rs1800796, (iv) IL-6-572G/C, (v) polymorphisms, (vi) GWAS, (vii) case–control study.

2.2 *Eligibility Criteria*

We explored the title and abstract for primarily removing the irrelevant studies. For the final database, we used the following inclusion–exclusion criteria. We included the study if: (i) the studies were designed to associate between IL-6 rs1800796 polymorphisms and cancer risk; (ii) the studies were case–control design; and (iii) the study provides suitable information about genotypic frequency. We excluded the study if: (i) not a case–control study; (ii) vague information is provided about genotypic distribution; and (iii) duplicated study.

2.3 *Data Extraction*

Using the eligible criteria, we compiled several information for each publication: first author, year of the study, country of origin, ethnicity of the study subject, number of case–control, types of cancer, and allelic and genotypic distribution. Additionally, Hardy–Weinberg equilibrium (HWE) test was performed based on control population to find out the quality of eligible studies for meta-analysis. The HWE test was quantified by the probability value (p-value) of Chi-square test with the null hypothesis that the study was consistent. Two authors independently lead the study search and data extraction. The eligible study for the meta data are, Bai et al. (2013), Bao et al. (2008), Chen et al. (2013, 2015), Dos Santos et al. (2018), Huang et al. (2016), Hwang et al. (2003), Kamangar et al. (2006), Kiyohara et al. (2014), Liang et al. (2013), Lim et al. (2011), Pierce et al. (2009), Seow et al. (2006), Slattery et al. (2007, 2009), Su and Zhou (2014), Tang et al. (2014), Tsilidis et al. (2009), Wang et al. (2009), Wang et al. (2018), Sun et al. (2004), Zhu et al. (2017).

2.4 *Statistical Analysis*

At first, we checked the quality of study through Hardy–Weinberg equilibrium (HWE), where the HWE test carried Chi-square test if $p < 0.05$ the required study was inconsistent. We used the Cochran's Q statistic and I2 to measure the heterogeneity of the eligible studies, where Q-test was performed based on Chi-square statistic. The p-value (<0.10) and I^2 (>50%) gives the consistency to estimate the pooled odds ratio (OR) by random effect model, otherwise fixed effect model. The pooled OR was assessed for checking the significant association between IL-6 polymorphisms and cancer risk under five genetic models: Dominant model (CC + CG vs. GG); Homozygote comparison (CC vs. GG); Over-dominant model (CG vs. CC + GG); Recessive model (CC vs. CG + GG); and Allele contrast (C vs. G). Beside the OR, we also estimated 95% confidence interval and compared using Z-test. Also, we generate the results according to the subgroup of ethnicity and types of cancer. The

HWE was followed by performing sensitivity analysis. All the statistical analysis was executed through "meta" package in R program.

3 Results

3.1 *Study Characteristics*

In this meta-analysis, we pulled together 150 studies preliminarily through titles and abstracts, and finally obtained 117 studies after removing any duplicates. Fifty-six articles are further removed for the reason of absence of full text, not case–control type, and not related with cancer diseases. Finally, 27 articles are included for analysis through complete information of the articles. The flow chart of the articles' selection process is shown in Fig. 1. The relevant articles have 27 studies of rs1800796 with 10,733 cases and 13,405 controls. The eligible studies have different types of cancer such as blood cancer, breast cancer, cervical cancer, colon cancer, liver cancer, lung

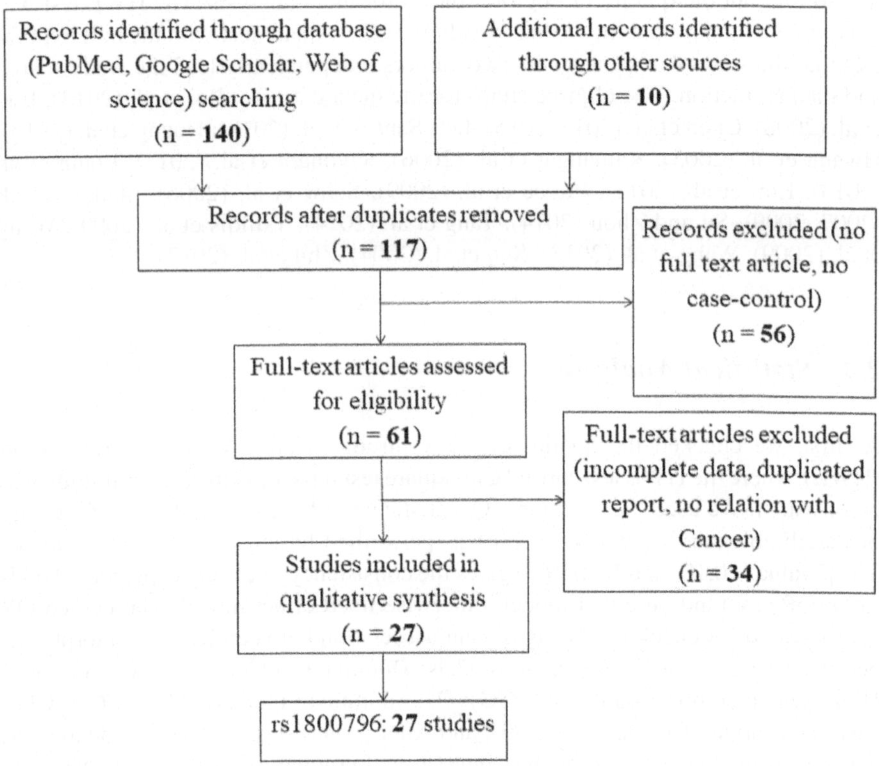

Fig. 1 Flow diagram of study selection for IL-6 gene rs1800796 (−572G/C) polymorphism; where "n" is the number of studies. *Source* created by the authors

cancer, neuroblastoma, oral cancer, ovarian cancer, pancreatic renal cell, prostate cancer, skin cancer, stomach cancer, and thyroid cancer. For being single studies, breast cancer and liver cancer for rs1800796 were organized in subgroup analyses as other cancer (Table 1) (to know about more information about the polymorphism, visit Harun-Or-Roshid et al. 2021).

3.2 *Quantitative Synthesis*

In accordance with IL-6-572G/C polymorphisms, we found insignificant association with overall cancer risk under four genetic models. There was a significant association found for over-dominant model (CG vs. CC + GG: OR = 1.12, 95% CI = 1.01 – 1.23, *p*-value = 0.03) of IL-6 572G/C polymorphism with increasing overall cancer risk (Table 1).

The subgroup analysis of IL-6-572G/C polymorphism expressed the significant association with defensive role in prostate cancer (C vs. G: OR = 0.74, 95% CI = 0.64–0.85, *p*-value = 0.00, (Fig. 2); CC vs. GG: OR = 0.52, 95% CI = 0.37 – 0.72, *p*-value = 0.00; CC vs. CG + GG: OR = 0.67, 95% CI = 0.53 – 0.84, *p*-value = 0.00; CC + CG vs. GG: OR = 0.74, 95% CI = 0.61 – 0.90, *p*-value = 0.00). The IL-6-572G/C also played significant association with increasing risk for lung cancer (CC + CG vs. GG: OR = 1.31, 95% CI = 1.04 – 1.65, *p*-value = 0.02; CG vs. CC + GG: OR = 1.31, 95% CI = 1.08 – 1.59, *p*-value = 0.01) and stomach cancer (C vs. G: OR = 1.16, 95% CI = 1.03–1.30, *p*-value = 0.01; CC vs. GG: OR = 1.41, 95% CI = 1.10 – 1.81, *p*-value = 0.01; CC vs. CG + GG: OR = 1.29, 95% CI = 1.07 – 1.55, *p*-value = 0.01). The colon cancer and other cancer showed no association with the polymorphism (Table 1) (for details, visit Harun-Or-Roshid et al. 2021).

3.3 *Source of Heterogeneity*

The significant heterogeneity were discerned in the analysis of IL-6-572G/C polymorphism for overall cancer risk (C vs. G: Q = 89.96, df = 26, *p*-value = 0.00, τ^2 = 0.0391, I^2 = 71.04; CC vs. GG: Q = 55.82, df = 26, *p*-value = 0.001, τ^2 = 0.10, I^2 = 53.41%; CC vs. CG + GG: Q = 49.23, df = 26, *p*-value = 0.004, τ^2 = 0.0459, I2 = 47.18%; CC + CG vs. GG: Q = 76.19, df = 26, *p*-value = 0.001, τ2 = 0.0618, I^2 = 65.88%; CG vs. CC + GG: Q = 54.40, df = 26, *p*-value = 0.001, τ^2 = 0.0293, I^2 = 52.21%; availability of results in Harun-Or-Roshid et al. 2021). For the subgroup analysis, we observed that the main sources of heterogeneity were found in colon, lung, and other cancer groups (Table 1).

Table 1 Summary of association and heterogeneity analysis of the IL-6 gene rs1800796 polymorphism with cancer risk

Cancer Type = Overall Cancers		Summary measures				Heterogeneity				
	Model	OR[a]	95% C.I.[b]	Z-value	p-value	Q-test	d.f	p-val	Tau^2	I^2
CC vs.GG	REM	1.03	[0.85; 1.25]	0.30	0.76	55.82	26	0.00	0.1033	53.41%
CC vs.CG + GG	REM	0.99	[0.86; 1.14]	−0.18	0.85	49.23	26	0.00	0.0459	47.18%
CC + CG vs.GG	REM	1.07	[0.94; 1.21]	1.05	0.29	76.19	26	0.00	0.0618	65.88%
CG vs.CC + GG	REM	**1.12**	**[1.01; 1.23]**	**2.19**	**0.03**	54.40	26	0.001	0.0293	52.21%
C vs.G	REM	1.04	[0.95; 1.15]	0.87	0.38	89.96	26	0.000	0.0391	71.07%
Cancer Type = Colon Cancers		Summary measures				Heterogeneity				
	Model	OR	95% C.I	Z-value	p-value	Q-test	d.f	p-val	Tau^2	I^2
CC vs.GG	REM	1.04	[0.67; 1.64]	0.19	0.85	6.75	2	0.03	0.0916	70.38%
CC vs.CG + GG	FEM	0.95	[0.82; 1.12]	−0.58	0.56	4.89	2	0.09	0.0532	59.12%
CC + CG vs.GG	FEM	0.97	[0.87; 1.10]	−0.44	0.66	9.87	2	0.01	0.0705	79.74%
CG vs.CC + GG	FEM	1.00	[0.89; 1.13]	0.00	0.99	3.47	2	0.18	0.0197	42.43%
C vs.G	REM	1.10	[0.79; 1.53]	0.58	0.56	13.42	2	0.001	0.0646	85.09%
Cancer Type = Lung Cancers		Summary measures				Heterogeneity				
	Model	OR	95% C.I	Z-value	p-value	Q-test	d.f	p-val	Tau^2	I^2
CC vs.GG	REM	1.13	[0.75; 1.69]	0.59	0.55	12.78	6	0.05	0.1477	53.00%
CC vs.CG + GG	FEM	0.93	[0.74; 1.17]	−0.62	0.54	10.58	6	0.10	0.0777	43.24%
CC + CG vs.GG	REM	**1.31**	**[1.04; 1.65]**	**2.28**	**0.02**	18.77	6	0.01	0.0604	68.04%
CG vs.CC + GG	REM	**1.31**	**[1.08; 1.59]**	**2.69**	**0.01**	14.24	6	0.03	0.0383	57.87%

(continued)

Table 1 (continued)

Cancer Type = Overall Cancers		Summary measures				Heterogeneity				
	Model	OR[a]	95% C.I.[b]	Z-value	p-value	Q-test	d.f	p-val	Tau^2	I^2
C vs.G	REM	1.19	[0.98; 1.43]	1.79	0.07	20.36	6	0.00	0.0423	70.53%
Cancer Type = Prostate Cancers		Summary measures				Heterogeneity				
	Model	OR	95% C.I	Z-value	p-value	Q-test	d.f	p-val	Tau^2	I^2
CC vs.GG	FEM	**0.52**	**[0.37; 0.72]**	**−3.91**	**0.00**	2.98	5	0.70	0.0000	0.00%
CC vs.CG + GG	FEM	**0.67**	**[0.53; 0.84]**	**−3.50**	**0.001**	4.44	5	0.49	0.0000	0.00%
CC + CG vs.GG	FEM	**0.74**	**[0.61; 0.90]**	**−3.04**	**0.00**	4.50	5	0.48	0.0000	0.00%
CG vs.CC + GG	FEM	1.00	[0.84; 1.18]	−0.01	0.98	4.64	5	0.46	0.0000	0.00%
C vs.G	FEM	**0.74**	**[0.64; 0.85]**	**−4.34**	**0.000**	5.23	5	0.39	0.0014	4.46%
Cancer Type = Stomach Cancers		Summary measures				Heterogeneity				
	Model	OR	95% C.I	Z-value	p-value	Q-test	d.f	p-val	Tau^2	I^2
CC vs.GG	FEM	**1.41**	**[1.10; 1.81]**	**2.67**	**0.01**	6.16	8	0.63	0.0000	0.00%
CC vs.CG + GG	FEM	**1.29**	**[1.07; 1.55]**	**2.65**	**0.00**	6.35	8	0.61	0.0000	0.00%
CC + CG vs.GG	FEM	1.14	[0.98; 1.34]	1.68	0.09	14.29	8	0.08	0.0510	44.00%
CG vs.CC + GG	REM	1.02	[0.79; 1.31]	0.13	0.89	18.79	8	0.02	0.0708	57.42%
C vs.G	FEM	**1.16**	**[1.03; 1.30]**	**2.70**	**0.00**	8.59	8	0.38	0.0023	6.92%
Cancer Type = Other Cancers		Summary measures				Heterogeneity				

(continued)

Table 1 (continued)

Cancer Type = Overall Cancers		Summary measures				Heterogeneity				
	Model	OR[a]	95% C.I.[b]	Z-value	p-value	Q-test	d.f	p-val	Tau^2	I^2
	Model	OR	95% C.I	Z-value	p-value	Q-test	d.f	p-val	Tau^2	I^2
CC vs.GG	REM	1.13	[0.47; 2.68]	0.27	0.79	4.51	1	0.03	0.3124	77.80%
CC vs.CG + GG	FEM	0.90	[0.78; 1.04]	−1.49	0.14	3.12	1	0.08	0.1651	67.96%
CC + CG vs.GG	REM	1.03	[0.64; 1.67]	0.14	0.89	5.31	1	0.02	0.0964	81.16%
CG vs.CC + GG	FEM	1.12	[0.99; 1.28]	1.73	0.08	0.44	1	0.51	0.0000	0.00%
C vs.G	REM	1.05	[0.72; 1.53]	0.27	0.79	7.94	1	0.01	0.0640	87.41%

[a]OR = Odds Ratio; [b]C.I. = Confidence Interval; REM = Random Effect Model; FEM = Fixed Effect Model. *Source* Authors' own calculation by using R)

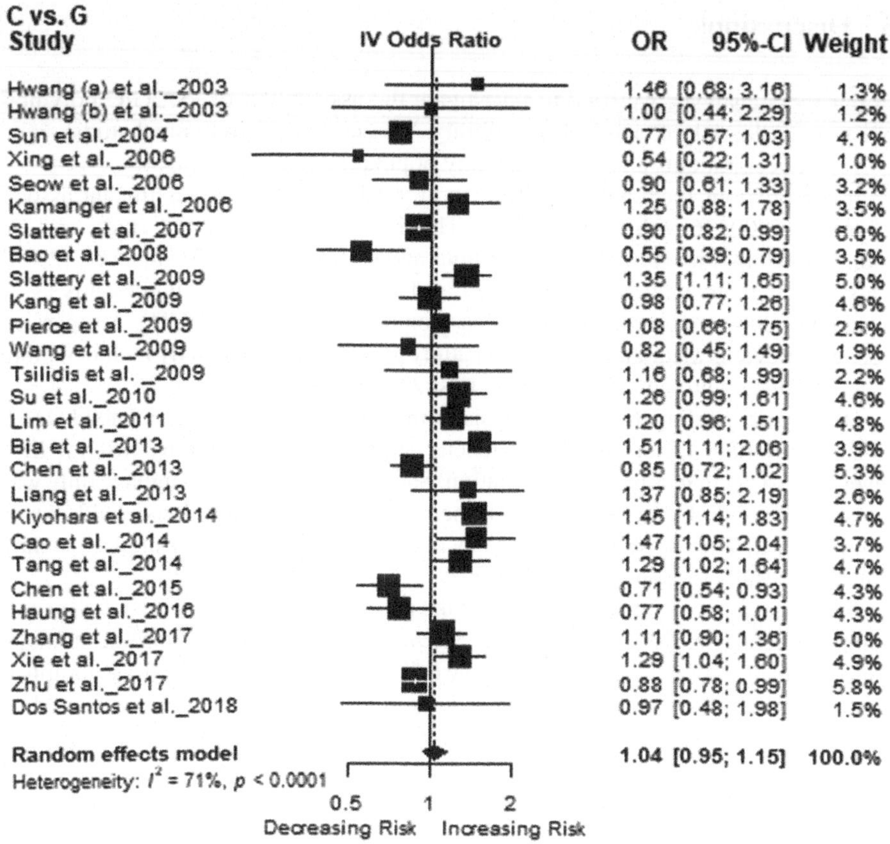

Fig. 2 Forest plot of IL-6-572G/C polymorphism with overall cancer risk under allelic model (C vs. G). The black square of the horizontal line represent the individual study-specific ORs with 95% Cis, and the area of the squares represent the corresponding study weight. The black diamond reflects the pooled OR, and the lateral points of the diamond represent the CI of the overall analyses. The solid vertical lines are the OR of 1 which is line of no effect. The dashed vertical line shows the corresponding pooled OR of the analyses. *Source* Created by the authors

4 Publication Bias

To check the publication bias of IL-6-572G/C polymorphism, we used the funnel plot with the allelic model C versus G, and we found the distribution of ORs in terms of standard errors (SEs) was symmetric. So there was no publication bias among the selected studies for meta-analysis. We also checked out the publication bias through Begg's test (C vs. G: p-value = 0.6022) and Egger's linear regression test (C vs. G: p-value = 0.3267), which suggested that there is no significant publication bias for this polymorphism (for further results available visit Harun-Or-Roshid et al. 2021).

5 Discussions

In our meta-analysis, we tried to investigate the association of -572G/C polymorphism with the risk of cancer susceptibility. We enrolled 27 individual studies with 10,733 cancer patients and 13,405 healthy population datasets.

In the previous meta-analysis, Zhou et al. (2018) showed the significant association of rs1800796 with overall cancer. This is also observed in our study, and we found more significant association for lung and stomach cancer. Wang et al. (2014) suggested in their study that IL-6-572G/C polymorphism was not significantly associated with lung cancer, but our study has contradicted their result. The comparison between previous meta-analysis and our study proposed that our study included more types of cancer and more sample sizes to stable the results.

However, regarding the meta-analysis, some limitations should be acknowledged. Firstly, important factors for heterogeneity such as age, sex, family history, IL-6 level, and so on, may affect the association. Secondly, the English language only was considered for collecting the datasets, therefore, the publication bias could not be completely avoided or selection bias may occur. Thirdly, the small sample size may affect the results for some types of cancer.

In conclusion, our study proposes that the people may have higher risk for overall cancers, decreasing risk for prostate cancer, and increasing risk for lung and stomach cancer, which carried C allele of IL-6 gene (-572G/C) polymorphism. To accomplish this research, it is evident from the collected information that the IL-6 gene could be treated as a prognostic biomarker for cancer treatment. However, for the validation of stable association, it would be necessary for an updated study with larger sample sizes and including all heterogeneous factors.

Conflicts of Interest The authors report no conflicts of interest.

Acknowledgements This work was supported in part by UGC Grant A-1416/5/52/R.U/science-11/18-19 from the University Grants Commission (UGC), Government of the People's Republic of Bangladesh. The author thanks an anonymous referee for making some useful suggestions in an earlier draft.

References

Bai, L., Yu, H., Wang, H., Su, H., Zhao, J., & Zhao, Y. (2013). Genetic single-nucleotide polymorphisms of inflammation-related factors associated with risk of lung cancer. *Medical Oncology, 30*(1), 414. https://doi.org/10.1007/s12032-012-0414-6. (Epub 2013 Jan 6; PubMed PMID: 23292870).

Bao, S., Yang, W., Zhou, S., & Ye, Z. (2008). Relationship between single nucleotide polymorphisms in −174G/C and −634C/G promoter region of interleukin-6 and prostate cancer. *Journal of Huazhong University of Science and Technology Medical Sciences, 28*, 693–696. (PubMed PMID: 19107369).

Chen, C. H., Gong, M., Yi, Q. T., & Guo, J. H. (2015). Role of interleukin-6 gene polymorphisms in the development of prostate cancer. *Genetics and Molecular Research, 14*(4), 13370–4. https://doi.org/10.4238/2015.October.26.34. (PubMed PMID: 26535651).

Chen, J., Liu, R. Y., Yang, L., Zhao, J., Zhao, X., Lu, D. et al. (2013). A two-SNP IL-6 promoter haplotype is associated with increased lung cancer risk. *Journal of cancer research and clinical oncology, 139*(2), 231–42. https://doi.org/10.1007/s00432-012-1314-z. (Epub 2012 Oct 2; PubMed PMID: 23052692; PubMed Central PMCID: PMC4535449).

Dos Santos, M. P., Sallas, M. L., Zapparoli, D., Orcini, W. A., Chen, E., Smith, M. D. A. C. et al. (2018). Lack of association between IL-6 polymorphisms and haplotypes with gastric cancer. *Journal of cellular biochemistry*. https://doi.org/10.1002/jcb.28220. ((Epub ahead of print) PubMed PMID: 30525242).

Gasche, J. A., Hoffmann, J., Boland, C. R., & Goel, A. (2011). Interleukin-6 promotes tumorigenesis by altering DNA methylation in oral cancer cells. *International Journal of Cancer, 129*(5), 1053–1063. https://doi.org/10.1002/ijc.25764.PMC3110561.PMID21710491

Harun-Or-Roshid,, M., Ali, M. B., Jesmin, & Mollah, M. N. H. (2021). Statistical meta-analysis to investigate the association between the Interleukin-6 (IL-6) gene polymorphisms and cancer risk. *PLoS One, 16*(3), e0247055. https://doi.org/10.1371/journal.pone.0247055. (PMID: 33684135; PMCID: PMC7939379).

Huang, W. J., Wu, L. J., Min, Z. C., Xu, L. T., Guo, C. M., Chen, Z. P. et al. (2016). Interleukin-6 -572G/C polymorphism and prostate cancer susceptibility. *Genetics and Molecular Research, 15*(3). https://doi.org/10.4238/gmr.15037563. (PubMed PMID: 27706719).

Hwang, I. R., Hsu, P. I., Peterson, L. E., Gutierrez, O., Kim, J. G., Graham, D.Y. et al. (2003). Interleukin-6 genetic polymorphisms are not related to Helicobacter pylori-associated gastroduodenal diseases. *Helicobacter, 8*(2), 142–8. (PubMed PMID: 12662382).

IACR-International Association of Cancer Research. (http://gco.iarc.fr/).

Kamangar, F., Abnet, C. C., Hutchinson, A. A., Newschaffer, C. J., Helzlsouer, K., Shugart, Y. Y., et al. (2006). Polymorphisms in inflammation-related genes and risk of gastric cancer (Finland). *Cancer Causes Control, 17*(1), 117–25. (PubMed PMID: 16411061).

Kiyohara, C., Horiuchi, T., Takayama, K., & Nakanishi, Y. (2014). Genetic polymorphisms involved in the inflammatory response and lung cancer risk: A case-control study in Japan. *Cytokine, 65*(1), 88–94. https://doi.org/10.1016/j.cyto.2013.09.015. (PubMed PMID: 24139238).

Kumari, N., Dwarakanath, B. S., Das, A., & Bhatt, A. N. (2016a). Role of interleukin-6 in cancer progression and therapeutic resistance. *Tumour Biology, 37*(9), 11553–11572. (Epub 2016 Jun 3; Review. PubMed PMID: 27260630).

Kumari, N., Dwarakanath, B. S., Das, A., & Bhatt, A. N. (2016b). Role of interleukin-6 in cancer progression and therapeutic resistance. *Tumour Biology, 37*(9), 11553–11572.

Li, J., Mo, H. Y., Xiong, G., Zhang, L., He, J., Huang, Z. F. et al. (2012). Tumor microenvironment macrophage inhibitory factor directs the accumulation of interleukin-17-producing tumor-infiltrating lymphocytes and predicts favorable survival in nasopharyngeal carcinoma patients. *The Journal of Biological Chemistry, 287*(42), 35484–35495.

Liang, J., Liu, X., Bi, Z., Yin, B., Xiao, J., Liu, H. et al. (2013). Relationship between gene polymorphisms of two cytokine genes (TNF-a and IL-6) and occurring of lung cancers in the ethnic group Han of China. *Molecular Biology Reports, 40*(2), 1541–6. https://doi.org/10.1007/s11033-012-2199-2. PubMed PMID: 23100065.

Lim, W. Y., Chen, Y., Ali, S. M., Chuah, K. L., Eng, P., Leong, S. S. et al. (2011). Polymorphisms in inflammatory pathway genes, host factors and lung cancer risk in Chinese female never-smokers. *Carcinogenesis, 32*(4), 522–9. https://doi.org/10.1093/carcin/bgr006. (PubMed PMID: 21252117).

Medscape.com. (2006). *Cancer patients typically have increased interleukin-6 levels*. American Society of Clinical Oncology 2006 Annual Meeting, Abstracts 8632 and 8633.

Miao, J. W., Liu, L. J., & Huang, J. (2014). Interleukin-6-induced epithelial-mesenchymal transition through signal transducer and activator of transcription 3 in human cervical carcinoma.

International Journal of Oncology, 45(1), 165–176. https://doi.org/10.3892/ijo.2014.2422.PMID24806843

Pierce, B. L., Biggs, M. L., DeCambre, M., Reiner, A. P., Li, C., Fitzpatrick, A. et al. (2009). C-reactive protein, interleukin-6, and prostate cancer risk in men aged 65 years and older. Cancer Causes Control, 20(7), 1193–203. https://doi.org/10.1007/s10552-009-9320-4. (Epub 2009 Mar 8; PubMed PMID: 19267250; PubMed Central PMCID: PMC2846958).

Seow, A., Ng, D. P., Choo, S., Eng, P., Poh, W. T., Ming, T. et al. (2006). Joint effect of asthma/atopy and an IL-6 gene polymorphism on lung cancer risk among lifetime non-smoking Chinese women. *Carcinogenesis, 27*, 1240–1244. (PubMed PMID: 16344268).

Slattery, M. L., Wolff, R. K., Herrick, J., Caan, B. J., & Samowitz, W. (2009). Tumor markers and rectal cancer: Support for an inflammation-related pathway. *International Journal of Cancer, 125*(7), 1698–704. https://doi.org/10.1002/ijc.24467. (PubMed PMID: 19452524; PubMed Central PMCID: PMC2768342).

Slattery, M. L., Wolff, R. K., Herrick, J. S., Caan, B. J., & Potter, J. D. (2007). IL6 genotypes and colon and rectal cancer. *Cancer Causes Control, 18*(10), 1095–105. (Epub 2007 Aug 13; PubMed PMID: 17694420; PubMed Central PMCID: PMC2442470).

Su, M., & Zhou, B. (2014). Association of genetic polymorphisms in IL-6 and IL-1ß gene with risk of lung cancer in female non-smokers. *Zhongguo Fei Ai Za Zhi, 17*(8), 612–7. https://doi.org/10.3779/j.issn.1009-3419.2014.08.06. PubMed PMID: 25130968.

Sun, J., Hedelin, M., Zheng, S. L., Adami, H. O., Bensen, J., Augustsson-Bälter, K., et al. (2004). Interleukin-6 sequence variants are not associated with prostate cancer risk. Cancer Epidemiology and Prevention Biomarkers, 3(10), 1677-1679. PubMed PMID: 15466986.

Tang, S., Yuan, Y., He, Y., Pan, D., Zhang, Y., Liu, Y. et al. (2014). Genetic polymorphism of interleukin-6 influences susceptibility to HBV-related hepatocellular carcinoma in a male Chinese Han population. *Human Immunology, 75*(4), 297–301. https://doi.org/10.1016/j.humimm.2014.02.006. (PubMed PMID: 24530755).

Tsilidis, K. K., Helzlsouer, K. J., Smith, M. W., Grinberg, V., Hoffman-Bolton, J., Clipp, S. L. et al. (2009). Association of common polymorphisms in IL10, and in other genes related to inflammatory response and obesity with colorectal cancer. *Cancer Causes Control, 20*(9), 1739–51. https://doi.org/10.1007/s10552-009-9427-7. (PubMed PMID: 19760027; PubMed Central PMCID: PMC4119174).

Vainer, N., Dehlendorff, C., & Johansen, J. S. (2018). Systematic literature review of IL-6 as a biomarker or treatment target in patients with gastric, bile duct, pancreatic and colorectal cancer. *Oncotarget, 9*(51), 29820–29841.

WHO, Cancer statistics. (2018). (https://www.who.int/news-room/fact-sheets/detail/cancer).

Wang, M. H., Helzlsouer, K. J., Smith, M. W., Hoffman-Bolton, J. A., Clipp, S. L., Grinberg, V. et al. (2009). Association of IL10 and other immune response- and obesity-related genes with prostate cancer in CLUE II. *Prostate, 69*(8), 874–85. https://doi.org/10.1002/pros.20933. (PubMed PMID: 19267370; PubMed Central PMCID: PMC3016874).

Wang, W., Chen, J., Zhao, F., Zhang, B., & Yu, H. (2014). Lack of association between a functional polymorphism (rs1800796) in the interleukin-6 gene promoter and lung cancer. *Diagnostic Pathology, 9*, 134. https://doi.org/10.1186/1746-1596-9-134. (PubMed PMID: 24984610; PubMed Central PMCID: PMC4100037).

Wang, X., Yang, F., Xu, G., & Zhong, S. (2018). The roles of IL-6, IL-8 and IL-10 gene polymorphisms in gastric cancer: A meta-analysis. *Cytokine, 111*, 230–236. https://doi.org/10.1016/j.cyto.2018.08.024. (Epub 2018 Sep 6; PubMed PMID: 30195978).

Xie, G., Yao, Q., Liu, Y., Du, S., Liu, A., Guo, Z. et al. (2012). IL-6-induced epithelial-mesenchymal transition promotes the generation of breast cancer stem-like cells analogous to mammosphere cultures. *International Journal of Oncology, 40*(4), 1171–1179. https://doi.org/10.3892/ijo.2011.1275.PMC3584811.PMID22134360

Zhang, J. Z., Liu, C. M., Peng, H. P., & Zhang, Y. (2017). Association of genetic variations in IL-6/IL-6R pathway genes with gastric cancer risk in a Chinese population. *Gene, 623*, 1–4. https://doi.org/10.1016/j.gene.2017.04.038. (PubMed PMID: 28442395).

Zhou, B., Shu, B., Yang, J., Liu, J., Xi, T., & Xing, Y. (2014). C-reactive protein, interleukin-6 and the risk of colorectal cancer: A meta-analysis. *Cancer Causes Control, 25*(10), 1397–405. https://doi.org/10.1007/s10552-014-0445-8. (Epub 2014 Jul 23; Review. PubMed PMID: 25053407).

Zhou, L., Zheng, Y., Tian, T., Liu, K., Wang, M., Lin, S. et al. (2018). Associations of interleukin-6 gene polymorphisms with cancer risk: Evidence based on 49,408 cancer cases and 61,790 controls. *Gene,, 670*, 136–147. https://doi.org/10.1016/j.gene.2018.05.104. (Epub 2018 May 26; PubMed PMID: 29842912).

Zhou, W., Zhang, S., Hu, Y., Na, J., Wang, N., Ma, X. et al. (2016). Meta-analysis of the associations between TNF-a or IL-6 gene polymorphisms and susceptibility to lung cancer. *European Journal of Medical Research, 20*, 28. https://doi.org/10.1186/s40001-015-0113-9. Retraction in: Zhou, W., Zhang, S., Hu, Y., Na, J., Wang, N., Ma, X. et al. (2016). *European Journal of Medical Research, 21*(1), 31. (PubMed PMID: 25889486; PubMed Central PMCID: PMC4438440).

Zhu, R. M., Lin, W., Zhang, W., Ren, J. T., Su, Y., He, J. R., et al. (2017). Modification effects of genetic polymorphisms in FTO, IL-6, and HSPD1 on the associations of diabetes with breast cancer risk and survival. *PLoS One, 12*, e0178850.

Two Level Logistic Regression Analysis of Factors Influencing Dual Form of Malnutrition in Mother–Child Pairs: A Household Study in Bangladesh

Md. Akhtaruzzaman Limon, Abu Sayed Md. Al Mamun, Kumkum Yeasmin, Md. Moidul Islam, and Md. Golam Hossain

Abstract Bangladesh is undergoing a nutrition transition associated with rapid social and economic transitions giving rise to the double burden of the malnutrition phenomenon. It is essential to investigate the household study of malnutrition among mother and under-five child pairs. The objective of this study was to determine the prevalence and risk factors of malnutrition among mother and under-five child pairs at the same household in Bangladesh. Secondary data from the BDHS-2014 was used in this study. The sample population of this study consisted of 7,368 married, currently non-pregnant Bangladeshi women with their under-five child. Descriptive statistics, Chi-square tests, and two-level binary logistic regression model were used in this study. The prevalence of underweight mother and under-five child pairs was 22.0%, and the prevalence of overweight mother and underweight child was near to 10%. It was found that only less than 20 percent (19.6%) mother and child pairs was found to be of normal weight (healthy). The two-level binary logistic model showed that division, type of residence, parents' education, household wealth index, mothers' age, and child birth weight are found to be risk factors of under nutrition among mother and under-five child pairs. Our selected model identified the risk factors of under nutrition among mother and under-five child pairs in Bangladesh. These factors can be considered for reducing the number of malnutrition among mother and under-five child pairs in Bangladesh.

Keywords Nutritional status · Multilevel logistic regression · Mother–child pair · Bangladesh

1 Introduction

Nutrition is the process of providing or obtaining the food necessary for health and growth and absorbing the nutrients in those foods (NHMRC 2012). Balanced diet is most important in achieving normal growth and development and for maintaining good health throughout life. The condition of the health of a person that results

Md. A. Limon · A. S. Md. Al Mamun · K. Yeasmin · Md. M. Islam · Md. G. Hossain (✉)
Department of Statistics, University of Rajshahi, Rajshahi, Bangladesh

B. K. Sinha et al. (eds.), *Data Science and SDGs*,
https://doi.org/10.1007/978-981-16-1919-9_4

due to the lack of one or more nutrients is called under nutrition. However, when there is an excess intake of nutrients, it results in over nutrition. Thus, the condition of malnutrition covers both the states of under nutrition and over nutrition. Overweight and underweight have long been recognized as two different public health problems, as underlying factors have been assumed separately (Barnett et al. 2011), but matter of anxiety is the paradoxical coexistence of child under nutrition and maternal overweight within the same household, often described as the "dual burden of malnutrition," which is a relatively a new phenomenon that has been described in studies from low- and middle-income countries including Benin, Brazil, China, Haiti, Guatemala, South Africa, Malaysia, and Mexico and recently Bangladesh as well (Steyn et al. 2011, 2012; Barquera et al. 2007).

Bangladesh is undergoing a nutrition transition associated with rapid social and economic transitions giving rise to the double burden of the malnutrition phenomenon (NIPORT 2014). The household wealth quintile in Bangladesh has been increasing during the last two decades. The changes in the composition of diet, usually accompanied with the changes in the level of physical activity referred to as nutritional transition and analyzed by Popkin (2003), is characterized as taking westernized diet high in meat, saturated fats, sugar, and energy with sedentary physical activities and increasing mental pressure instead of a traditional, homemade food, low fat, high-fiber, plant-based diet combined with physical labor. Rather these, populations in many developing countries like Bangladesh are consuming more processed foods, including more refined grains and foods with higher content of saturated fat, sugar, and salt, with increasing economic development and urbanization (Doak et al. 2005). The nutritional transition usually occurs in parallel with economic, epidemiological, and demographic transitions in a country with a sedentary occupation and habituated with daily household labor saving technologies.

The number of overweight and obese women has been unrelenting to increase internationally including in low-income countries (WHO 2011) like Bangladesh (Hossain et al. 2012), on the other hand under nutrition is associated with one-third to one-half of the deaths of child under five years of age globally (Black 2008). In sub-Saharan Africa, 28% of child under five years of age are moderately or severely underweight (UNICEF 2007) with 38% of child under five stunted (UNICEF 2007). Meanwhile, global obesity prevalence has doubled since 1980 (WHO 2011). In a study of several African countries (Burkina Faso, Ghana, Kenya, Malawi, Niger, Senegal, and Tanzania), the prevalence of urban adult overweight and obesity increased around 35% from 1992 to 2005 (Ziraba 2009). In a more recent study using Demographic Health Survey data from the early 2000s from North Africa, sub-Saharan Africa, Asia, and Latin America, it had been found that low levels of maternal education, working in agriculture, living in urban areas, increased siblings in the household, and relative poverty were associated with increased risk of dual burden households (Wojcicki 2014; Jehn and Brewis 2009). From a lot of literature review (Sumon et al. 2020; Hossain et al. 2018; Hossain et al. 2012; etc.), it has been examined that there is no survey about the dual burden of malnutrition (experiencing under nutrition and over nutrition in the same household) of Bangladeshi population.

The aim and objectives of this study was to determinate the prevalence and associated factors of malnutrition among mother and under-five child pairs in Bangladesh.

2 Methods

This cross-sectional study extracted data from Bangladesh Demographic and Health Survey (BDHS-2014). BDHS-2014 collected data from overall Bangladesh using two-stage stratification procedure. The sampling technique, survey design, survey instruments, measuring system, and quality control have been described elsewhere (NIPORT 2014). If a mother had more than one under-five children, we considered the last born child for this study.

2.1 Outcome Variable

The nutritional status of mother and their under-five child was the outcome variable in this study. Nutritional status was measured by the body mass index (BMI) for mother and for child by weight-for-age z-score. Nutritional status of mothers was classified into three classes according to cut off points of BMI:

(i) Under nutrition (BMI < 18.5 kg/m^2),
(ii) Normal weight ($18.5 \leq$ BMI < 25 kg/m^2), and (iii) over nutrition (BMI $\geq$ 25 kg/m^2).

There were three categories of the outcome variable. The principal concern of this study was the categories,

(i) Under nourished mother–child pairs.
(ii) Over nourished mother–child pairs.
(iii) Normal weight mother–child pairs.

But the frequency of over nourished mother and over nourished child was very small (0.2%), finally we considered only under nourished mother–child pairs (Code, 1) compared to normal weight of mother–child pairs (Code, 0).

2.2 Independent Variables

Socioeconomic, demographic, and household information were considered as independent variables in our study that came from the relevant record of 2014-BDHS survey.

2.3 Statistical Analysis

Descriptive statistics (frequency distribution) was used to determine the prevalence of nutritional status of mothers and their under-five child. Chi-square (χ^2) tests were performed to find the association between dependent and independent variables. The significantly associated factors provided by χ^2-test were used as independent variables in two-level logistic models. Sample was selected from different level, a traditional (single-level) statistical model is not appropriate for the analysis of such data (Khan and Shaw 2011). Two-level binary logistic regression analysis was used to examine the effect of selected independent variables on our outcome variable. Statistical significance was accepted at $p < 0.05$. All statistical analyses were performed using SPSS (IBM version 20.0) Software.

3 Results

A total of 7,368 married, currently non-pregnant Bangladeshi women and their under-five child were selected for the study. The age range of mothers was from 15 to 49 years with an average age of 31.52 ± 9.17 years.

3.1 Prevalence of Nutritional Status Among Mother–Child Pairs

It was noted that more than one-fifth (22.0%) mother–child pairs were under nourished while the prevalence of overweight mother and underweight child were near to 10%. It was found that 19.6% mother and child pair was normal weight (healthy) (Table 1).

χ^2-test demonstrated that child birth weight, type of residence, fathers' education level, geographical location (division), mothers' educational level, household wealth status, mothers' occupation, mothers' age, place of delivery, and mothers' anemia were significantly associated factors of under nutrition among mother–child pairs

Table 1 Distribution of nutritional status among mother–child pairs in Bangladesh

Mother–child pairs' nutritional status	N (%)
Under nutrition among mother–child pairs	1621 (22.0%)
Over nourished mothers and under nourished child	710 (9.6%)
Normal weight mother–child pairs	1441 (19.6%)

Source Authors' own calculation by using SPSS

in Bangladesh. These factors were considered as independent variables in two-level binary logistic regression model.

3.2 Two-Level Binary Logistic Regression Model

It was observed that the standard error (SE) of each independent variable was between values of 0.001 and 0.5, there was no evidence of multicollinearity problem. The binary logistic model demonstrated that after controlling the effect of other variables, the mothers and child living in Sylhet division were more likely to be both underweight than those were living in Rangpur [AOR = 0.640, 95% CI: 0.417–0.983; $p < 0.01$], Barisal [AOR = 0.599, 95% CI: 0.380–0.944; $p < 0.01$], Rajshahi [AOR = 0.480, 95% CI: 0.308–0.746; $p < 0.01$], Dhaka [AOR = 0.475, 95% CI: 0.315–0.717; $p < 0.01$], Chittagong [AOR = 0.449, 95% CI: 0.298–0.676; $p < 0.01$], and Khulna [AOR = 0.312, 95% CI: 0.201–0.485; $p < 0.01$] divisions. It was found that rural mothers and their under-five child had more chance to get underweight than urban mothers and their under-five child (AOR = 1.97, CI: 1.268–3.076; $p < 0.01$). Education level of mother was found to have significant effect on UMUC. Mothers with no education (AOR = 3.14, CI: 1.91–5.15; $p < 0.01$), primary education (AOR = 2.85, CI: 1.799–4.54; $p < 0.01$), secondary education (AOR = 1.71, CI: 1.109–2.64; $p < 0.01$) had more chance of occurring underweight mother–child pair than higher educated mothers. Highly educated fathers had played a significant role in their family's nutritional status. This study showed that uneducated (AOR = 3.96, CI: 2.70–5.78; $p < 0.01$), and primary (AOR = 3.11, CI: 2.17–4.47; $p < 0.01$) and secondary educated (AOR = 2.19, CI: 1.55–3.10; $p < 0.01$) husbands' wives and their under-five child both were more likely to get underweight than higher educated husbands' wives and their under-five child. From Table 2, we observed that family's wealth status has significant effect on mother–child pairs' (UMUC) nutritional status. The odds ratios provide that the risk of arising underweight mother and underweight child among less than 20 years aged mothers is 0.676 and 0.876 times higher than 21–29 years (AOR = 0.676, CI: 1.55–3.10; $p < 0.01$) and above 30 years aged mothers (AOR = 0.876, CI: 1.55–3.10; $p < 0.01$), respectively. The adjusted odds ratio of child birth weight (smaller than average vs. average) is 2.55 [(95% CI: 1.579, 4.131; $p < 0.038$)], which indicates that the risk of getting underweight mother and underweight child was 2.55 times higher when the child had low birth weight than the average birth weight child. The adjusted odds ratio reveals that the poor are found to have 0.62 [(95% CI: 1.124, 2.067); $p < 0.001$] times and 0.388 [(95% CI: 0.237, 0.633); $p < 0.001$] times higher risk of underweight mother and underweight child than the middle class and rich people. The Nagelkerke R square value showed that the multiple binary logistic regression models explained 49.2% of the variation in the outcome variable (UMUC). Hosmer and Lemeshow test also proved the good fitting of the model to the data (Table 2).

Table 2 Two-level logistic regression estimates for the effects of demographic and socioeconomic factors on under nutrition of mother–child pairs

Variable	B	SE	Wald	P-value	AOR	95% CI for AOR	
						Lower	Upper
Division			33.62	$p < 0.001$			
Chittagong vs. Sylhet[R]	−0. 79	0.20	9.65	$p < 0.002$	0.44	0.29	0.67
Dhaka vs. Sylhet[R]	−0.74	0.20	18.00	$p < 0.001$	0.47	0.31	0.71
Khulna vs. Sylhet[R]	−1.16	0.22	15.86	$p < 0.001$	0.31	0.20	0.48
Rajshahi vs. Barisal[R]	−0.73	0.22	28.70	$p < 0.001$	0.48	0.30	0.74
Rangpur vs. Sylhet[R]	−0.44	0.21	14.90	$p < 0.001$	0.64	0.41	0.98
Barisal vs. Sylhet[R]	−0.51	0.23	15.61	$p < 0.001$	0.59	0.38	0.94
Type of residence			23.67	$p < 0.002$			
Rural vs. urban[R]	0.68	0.22	3.01	$p < 0.003$	1.97	1.26	3.07
Mothers' education level			13.64	$p < 0.003$			
Secondary vs. higher [R]	0.53	0.22	2.43	$p < 0.035$	1.71	1.10	2.64
Primary vs. higher [R]	1.05	0.23	4.45	$p < 0.035$	2.85	1.79	4.54
No education vs. higher[R]	1.14	0.25	4.54	0.488	3.14	1.91	5.15
Husbands' education level			16.97	$p < 0.001$			
Secondary vs. higher[R]	0.78	0.17	4.45	$p < 0.001$	2.19	1.55	3.10
Primary vs. higher[R]	1.13	0.18	6.18	$p < 0.001$	3.11	2.17	4.47
No education vs. higher[R]	1.37	0.19	7.10	$p < 0.009$	3.96	2.70	5.78
Mother's age group			9.98	$p < 0.007$			

(continued)

Table 2 (continued)

Variable	B	SE	Wald	P-value	AOR	95% CI for AOR	
						Lower	Upper
21–29 vs. ≤ 20[R]	−0.39	0.35	−1.85	$p < 0.005$	0.67	0.44	1.02
≥ 30 vs. ≤ 20[R]	−0.13	0.27	−0.48	0.467	0.87	0.51	1.49
Child birth weight			12.66	$p < 0.003$			
Low vs. average[R]	0.93	0.24	3.82	$p < 0.001$	2.55	1.57	4. 13
Household wealth status			71.79	$p < 0.001$			
Middle vs. poor[R]	−0.47	0.23	−2.03	$p < 0.001$	0.62	0.39	0.98
Rich vs. poor[R]	−0.94	0.25	−3.78	$p < 0.001$	0.38	0.23	0.63
Constant	−2.07	0.31	45.10	0.000	0.12		
Goodness of fit	Hosmer and Lemeshow Test			P value = 0.43			

Source Authors' own calculation by using R. N.B: B = Co-efficient, SE = Standard Error, AOR = Adjusted Odds ratio, CI = Confidence Interval

4 Discussion

4.1 Prevalence of Under Nutrition

In our study, it was found that the prevalence of underweight mother and underweight child (UMUC) pair in Bangladesh was 22.0%. As far as we were concerned, no other studies were available regarding nutritional status of mother and their under-five child pairs in Bangladesh. Some studies have been done with mothers and their under-five child nutritional status, but none of the studies included Bangladesh (Wong et al. 2015; Doak et al. 2005).

4.2 Effect of Socioeconomic and Demographic Factors on Under Nutrition

It was observed that the prevalence of underweight mothers and their underweight child among the divisions varied due to different culture and different level of socioeconomic factors. The result of this study showed that prevalence of underweight mother and underweight child was higher among rural mother and child. The socioeconomic facilities of Bangladesh are not equivalently enabled among rural and urban areas. In rural areas, there are some obvious deficiencies in many socioeconomic conveniences such as nutritional, medical, and educational facilities. Mother and child living in rural Bangladesh do not get the same socioeconomic facilities as urban mother and child. This might be responsible for the higher percentage of underweight mother and child pair in rural areas compared to urban areas. This study also showed that comparatively lower educated mothers are more likely to be underweight with their under-five child in Bangladesh. This study reveals that the highest and the lowest number of underweight mother and child pair exist among illiterate mothers (67.8%) and higher educated mothers (22.6%). Education plays a pivotal role in maintaining good health. With no education, mothers do not have the knowledge of nutrition and health. Consequently, they remain unaware of the nutritional status of themselves and child as well. Higher education empowers a woman to acquire knowledge about good health and nutrition. More number of uneducated and primary educated mothers with poor family assets are living in rural environment than urban area. These might be possible causes for getting underweight among lower educated mothers and their under-five child.

5 Conclusion

A total of 7,368 mother and their under-five child pairs were considered as the sample in the present study. In the study, we found that a remarkable number of mother–child pairs in Bangladesh were suffering from under nutrition. We observed that some modifiable factors such as household wealth index, mother's education, child's birth weight, and type of residence have impact on under nutrition of mother–child pairs. These factors can be considered for reducing the number of malnutrition among mothers and their under-five child in Bangladesh.

6 Limitation of the Study

In this study, we used secondary data, and it was bounded by the limitations of those data. The foremost limitation of this study is that it was a cross-sectional study. Since the study was a cross-sectional study, it was difficult to set up a causal relationship between the socioeconomic, demographic, and anthropometric factors and mother–child pair's nutritional status in Bangladesh. In this study, we measured the child's nutritional status by using only weight-for-age z-score. There are two more indices of measuring child nutritional status, height for-age (stunting) and weight-for-height (wasting), which we did not use in our study. Clearly, more research is required regarding the other paradoxical forms of mother–child pair's malnutrition in Bangladesh.

Acknowledgements We would like to thank the Bangladesh Demographic and Health Survey (BDHS) for providing nationally representative-based data collected in 2014.

References

Barnett, I. (2011). *Is the dual burden of over- and undernutrition a concern for poor households in Ethiopia, India, Peru and Vietnam?* Young Lives.

Barquera, S., Peterson, K. E., Must, A., et al. (2007). Coexistence of maternal central aadiposity and child stunting in Mexico. *International Journal of Obesity, 31*, 601–607.

Black, R. E., Allen, L. H., Bhutta, Z. A., et al. (2008). Maternal and child undernutrition: Global and regional exposures and health outcomes. *Lancet, 371*(9608), 243–260.

Doak, C. M., Adair, L. S., Bentley, M., et al. (2005). The dual burden household and the nutrition transition paradox. *International Journal of Obesity, 29*(1), 129–136.

Hossain, M. G., Bharati, P., Aik, S. A., et al. (2012). Body mass index of married Bangladeshi women: Trends and association with socio-demographic factors. *Journal of Biosocial Science, 44*(4), 385.

Hossain, M., Islam, A., Kamarul, T., & Hossain, G. (2018). Exclusive breastfeeding practice during first six months of an infant's life in Bangladesh: A country based cross-sectional study. *BMC Pediatrics, 18*(1), 1–9.

Jehn, M., & Brewis, A. (2009). Paradoxical malnutrition and the phenomenon of over and undernutrition in underdeveloped economies. *Economics & Human Biology, 7*, 28–35.

Khan, M. H. R., & Shaw, J. E. H. (2011). Multilevel logistic regression analysis applied to binary contraceptive prevalence data. *Journal of Data Science, 9*, 93–110.

NHMRC: National Health and Medical Research Council. Australian Dietary Guidelines 2012. https://www.nhmrc.gov.au/health-topics/nutrition.

NIPORT: National Institute of Population Research and Training, Mitra and Associates, ICF International (2014) Bangladesh Demographic and Health Survey, 2014. NIPORT, Mitra & Associates and ICF International, Dhaka, Bangladesh and Calverton, MD, USA.

Popkin, B. M. (2003). The nutrition transition in the developing world. *Development Policy Review, 21*(5–6), 581–597.

Steyn, N. P., Labadarios, D., Nel, J. H., et al. (2011). What is the nutritional status of children of obese mothers in South Africa? *Nutrition, 27*(9), 904–911.

Steyn, N. P., Nel, J. H., Parker, W., et al. (2012). Urbanisation and the nutrition transition: A comparison of diet and weight status of South African and Kenyan women. *Scandinavian Journal Public Health, 40*, 229–238.

Sumon, K., Sayem, M. A., Al Mamun, A. S. M., et al. (2020). Factors influencing in early childbearing and its consequences on nutritional sstatus of Bangladeshi mothers: Nationally representative data. *Research Square (preprint)*. https://doi.org/10.21203/rs.3.rs-67538/v1

UNICEF 2007: Progress for children: A world fit for children statistical review, number 6. New York: United Nation's Children Fund 2007.

WHO: Obesity and Overweight (Fact Sheet No. 311). Geneva: WHO 2011.
Wojcicki, J. M. (2014). The double burden household in sub-Saharan Africa: Maternal overweight and obesity and childhood undernutrition from the year 2000: Results from World Health Organization Data (WHO) and Demographic Health Surveys (DHS). *BMC Public Health,14*(1), 1–2.
Wong, C., Odom, S. L., Hume, K. A., et al. (2015). Evidence-based practices for children, youth, and young adults with autism spectrum disorder: A comprehensive review. *Journal of Autism and Development Disorders, 45*(7), 1951–1966.
Ziraba, A. K., Fotso, J. C., & Ochako, R. (2009). Overweight and obesity in urban Africa: A problem of the rich or the poor? *BMC Public Health, 9*, 465.

Divide and Recombine Approach for Analysis of Failure Data Using Parametric Regression Model

Md. Razanmiah, Md. Kamrul Islam, and Md. Rezaul Karim

Abstract The failure data of some products depend on factors or covariates such as the operating environment, usage conditions, etc. Under this situation, the parametric regression model is applied for modeling the failure data of the product as a function of covariates. Divide and recombine (D&R) is a new statistical approach to the analysis of big data. In the D&R approach, the data are divided into manageable subsets, an analytic method is applied independently to each subset, and the outputs are recombined. This chapter applies the D&R approach for analysis of an automobile component failure data using the Weibull regression model. Extensive simulation studies are presented to evaluate the performance of the proposed methodology with comparison to the traditional statistical estimation method.

Keywords Big data · Divide and recombine (D&R) approach · Failure data · Weibull regression model

1 Introduction

Automotive manufacturing companies analyze field reliability data to enhance the quality and reliability of their products and to improve customer satisfaction. In recent years, many manufacturers utilize the warranty database as a prime source of field reliability data, which can be collected economically and efficiently through repair service networks. Warranty claim data is superior to laboratory test data in the sense that it contains information on the actual environment in which the product is used (Karim and Suzuki 2007). Therefore, several procedures have been developed for collecting and analyzing warranty claim data, e.g., (Blischke et al. 2011) and the references given therein.

In this chapter, we discuss an approach for modeling the reliability of a specific system (unit) of automotive components based on field failure warranty data. The

Sections of the chapter draw from the co-author's (Md. Rezaul Karim) previous published works, reused here with permissions (Karim and Islam 2019) and (Karim and Suzuki 2006).

Md. Razanmiah · Md. K. Islam · Md. R. Karim (✉)
Department of Statistics, University of Rajshahi, Rajshahi, Bangladesh

B. K. Sinha et al. (eds.), *Data Science and SDGs*,
https://doi.org/10.1007/978-981-16-1919-9_5

unit's lifetime depends on some explanatory variables or covariates such as the automobile operating environment or used region, types of automobiles that use the unit, and the types of failure modes. If a unit fails within the warranty period, the information on covariates can be known from the warranty database.

This chapter applies the D&R approach for analysis of an automobile component failure data using the Weibull regression model. Extensive simulation studies are presented to evaluate the performance of the proposed methodology for the case of big data with comparison to the traditional statistical estimation method.

2 Parametric Regression Model

Regression analysis of lifetimes involves specifications for the distribution $f(t|X, \beta, \sigma)$ of a lifetime variable, T, given a vector of explanatory variables or covariates, $X = (x_1, \ldots, x_p)$, upon which lifetime may depend. Let $\theta = (\beta', \sigma, \gamma')'$ be a $(p+r+1) \times 1$ vector of all parameters in the model, where $\beta = (\beta_1, \ldots, \beta_p)'$ represents a vector of regression coefficients, σ is a scale parameter, and $\gamma = (\gamma_1, \ldots, \gamma_r)'$ is the parameter vector associated with the distribution of covariates X, $g(X|\gamma)$. The complete-data log-likelihood function based on n independent observations can be written as

$$l_{X,t}(\theta|X, t) = \sum_{i=1}^{n} l_{X,t}(\theta|X_i, t_i) = \sum_{i=1}^{n} l_{t|X}(\beta, \sigma|X_i, t_i) = \sum_{i=1}^{n} l_X(\gamma|X_i), \quad (1)$$

where $l_{X,t}(\theta|X_i, t_i)$ is the complete-data log-likelihood of θ for the ith observation based on the joint distribution of (X, t); $l_{t|X}(\beta, \sigma|X_i, t_i)$ is the log-likelihood based on the conditional distribution of $t|X$; and $l_X(\gamma|X_i)$ is the contribution from the marginal distribution of X (Karim and Suzuki 2006) and (Karim and Suzuki 2007). Here, we consider Weibull distribution for $f(t|X, \beta, \sigma)$. We assume the Weibull regression model in which the log-lifetime $Y = \log(T)$ follows a location-scale distribution with location parameter dependent on X, $\mu(X) = \beta' X$, and scale parameter σ. Under this model, the density function of T given X can be written as

$$f(t|X, \beta, \sigma) = \frac{1}{\sigma t} \exp\left[\left(\frac{\log(t) - \mu(X)}{\sigma}\right) - \exp\left(\frac{\log(t) - \mu(X)}{\sigma}\right)\right], \quad t > 0. \quad (2)$$

For more detailed explanations of Weibull regression model, see (Meeker and Escobar 1998; Lawless 2003; Karim and Suzuki 2007; Blischke et al. 2011; Karim and Islam 2019).

3 Divide and Recombine

The traditional estimation procedures for the parameters of the model (2) are quite attractive if the size of the sample is small, moderate, or large in the usual statistical sense. However, in the case of big data, it would not be possible to use the same procedure due to the amount of data being captured and stored. We need to review the theory, methodology, and computation techniques for big data analysis. Let us consider a situation where we have to consider 1,000,000 observations with 100 variables including both outcome and explanatory variables for each item resulting in a total of $1{,}000{,}000 \times 100$ observations. In reality, the data would be much more complex and bigger. This is not a typical statistical challenge simply due to the size of data, and hence, we need to find a valid way to use all the data without sacrificing statistical rigor. In this case, one logical solution is to divide and recombine data, see (Guha et al. 2012; Cleveland and Hafen 2014; Hafen 2016; Liu and Li 2018).

The idea is simple, we have to divide the big data into subsets, each analytic method is applied to subsets, and the outputs are recombined in a statistically valid manner. In the process of dividing and recombining, the big data set is partitioned into manageable subsets of smaller data, and analytic methods such as fitting of models are performed independently for each subset. One way to recombine is to use the average of the estimated model coefficients obtained from each subset (Guha et al. 2012). The resulting estimates may not be exact due to the choice of the recombining procedure but statistically valid. The advantage is obvious, we can make use of statistical techniques without any constraint arising from big data using R or available statistical packages.

(Lee et al. 2017) summarized D&R steps as follows: (i) the subsets are obtained by dividing the original big data into manageable smaller groups; (ii) the estimates or sufficient statistics are obtained for the subsets; and (iii) the results from subsets are combined by using some kind of averaging to obtain the estimate for the whole data set. According to (Hafen 2016), the division into subsets can be performed by either replicate division or conditioning variable division. Replicate division takes into account random sampling without replacement, and the conditioning variable division considers stratification of the data based on one or more variables included in the data. A feasible measure of a good fit is the least discrepancy with the estimate obtained from the entire data set. Other than a few exceptions, D&R results are approximate (Lee et al. 2017).

The steps for D&R are discussed below and Fig. 1 illustrates the computational procedure of the parametric regression model.

Step I: Divide the data into S subsets of similar structure, with T_s and X_s being the vector of responses and covariate matrix in subset s ($s = 1,2,\ldots, S$).

Step II: For the sth partitioned subset ($s = 1,2,\ldots,S$), compute $\hat{\theta}_s, s = 1, \ldots, S$, solving the maximum likelihood estimating equations derived based on Eqs. (1) and (2). The R function "survreg()" can be used for estimating $\theta_s, s = 1, \ldots, S$.

Step III: Recombine using the estimates obtained in Step II as follows

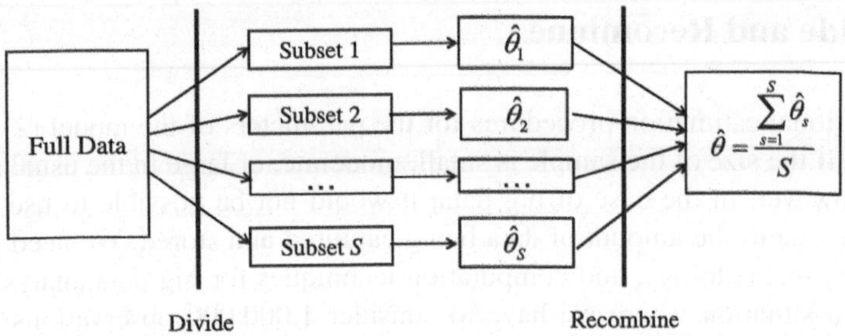

Fig. 1 Flowchart displaying D&R method for parametric regression model. *Source* Created by the authors

$$\hat{\theta} = \frac{\sum_{s=1}^{S} \hat{\theta}_s}{S}. \tag{3}$$

4 Example

We consider for illustration a set of failure (warranty claims) data for a specific system (unit) of an automobile. The units were sold during 26 months and warranty claims were recorded during 30 months observational period under the warranty of 18 months. The examination of the originally recorded data revealed a few types of typographical errors. After editing the relevant errors, the edited data were considered for analysis. The data set includes 4776 observations with the variables date of sale, date of claim, age, mileage, failure mode, unit's used region, etc. The information regarding the names of the unit, failure modes, and used regions are not disclosed here to protect the proprietary nature of the information.

Our interest is to investigate how the usage-based lifetime (mileage) of the unit differs with respect to age in days (X_1) and two categorical covariates: used region [Region (X_2): Region1 = 1, Region2 = 2, Region3 = 3, Region4 = 4], and failure modes [Mode (X_3): Mode1 = 1, Mode2 = 2, Mode3 = 3]. We assume a Weibull model for usage $T, f(t|X, \beta, \sigma)$, with location parameter dependent on covariates X, $\mu(X) = \beta' X = \beta_0 + \beta_1 X_1 + \beta_2 X_2 + \beta_3 X_3$, and scale parameter σ.

Tables 1, 2, and 3 summarize the numerical results for the Weibull regression model. Tables 1 and 2 show the MLEs of the parameters using all data at a time and the MLEs of the same parameters obtained by D&R method (with 10 random divisions), respectively.

The likelihood ratio chi-square statistic for all data taken at a time shows the p-value < 2e-16, indicates the null hypothesis that all regression parameters are zero is rejected. Under the divine and combined method, the parameters are estimated based on $S = 10$ random divisions. The estimated parameters obtained by the divine and combined method are approximately the same as that obtained by taking all the data at a time.

Table 1 MLEs of parameters using all data at a time (automobile data)

Coefficients	MLEs	Std. Error	95% Confidence intervals	
			Lower limit	Upper limit
Intercept (β_0)	9.658761	0.025215	9.609340	9.708181
Age (in days) (β_1)	0.002036	0.000064	0.001912	0.002161
Region2 (β_2)	−0.239107	0.026639	−0.291318	−0.186895
Region3 (β_3)	−0.244266	0.039409	−0.321506	−0.167027
Region4 (β_4)	−0.169698	0.019092	−0.207117	−0.132280
FM2 (β_5)	0.031099	0.019276	−0.006681	0.068880
FM3 (β_6)	0.005599	0.020943	−0.035448	0.046646

Source Authors' own calculation by using R

Table 2 MLEs of parameters through D&R method with 10 random divisions (automobile data)

Coefficients	MLEs	Std. Error	95% Confidence intervals	
			Lower limit	Upper limit
Intercept (β_0)	9.629127	0.079138	9.474020	9.784235
Age (in days) (β_1)	0.002077	0.000198	0.001688	0.002466
Region2 (β_2)	−0.192284	0.083816	−0.356560	−0.028008
Region3 (β_3)	−0.236043	0.117568	−0.466473	−0.005614
Region4 (β_4)	−0.157616	0.059221	−0.273687	−0.041545
FM2 (β_5)	0.036539	0.059934	−0.080930	0.154007
FM3 (β_6)	0.015934	0.064722	−0.110919	0.142787

Source Authors' own calculation by using R

Table 3 Comparison of MLEs of percentiles for age = 365 days, region = 1, and mode = 1

Percentile	Using all data at a time	Through the D&R method (10 random divisions)
10th percentile	9187.45	9108.55
25th percentile	16,242.09	16,055.05
50th percentile	26,747.01	26,393.65
75th percentile	39,630.50	39,076.22

Source Authors' own calculation by using R

Table 3 compares the MLEs of the percentiles (10th, 25th, 50th, and 75th) obtained by two methods for given covariate values, age = 365 days, used region1, and failure mode1. This table indicates that the percentiles estimated by two methods are approximately similar.

5 Simulation Study

To investigate the performance of the D&R approach, here we conduct a simulation study. In this simulation, we consider the true values of the parameters $\beta = (9.65, 0.002, -0.23, -0.24, -0.16, 0.03, 0.005)$ as the coefficients corresponding to the covariates $X = (1, X_A, X_{R2}, X_{R3}, X_{R4}, X_{FM2}, X_{FM3})$. The generated observations, $n = 100{,}000$, number of division $S = 100$, and number of simulations is 100. The summarized simulation results for different sizes of samples (200, 400, 600, and 800) are shown in Table 4.

Table 4 indicates that if the sample size increases, the estimates of the D&R approach becomes closer to the estimates obtained by the full data set. This implies the applicability of the D&R approach for analyzing failure data using the Weibull parametric regression model.

Figure 2 shows the root mean square error (RMSE) of the MLEs of parameters obtained from the simulation study. This figure indicates that the RMSE values decreased substantially with increasing sample sizes. It implies that the D&R method is practically useful in this case.

6 Conclusion

This chapter applied the Divide and Recombine approach for analysis of failure data. A set of warranty claims data of an automobile component is considered as an example. Weibull parametric regression model is assumed for the data, and the maximum likelihood estimation method is used for estimating the parameters of the model. Simulation studies are performed for investigating the performance of the approach. It is observed from the simulation results that the proposed method is practically useful. An extension of the method concerning more lifetime distributions would be valuable. Also, the automobile component that is considered in the example section has censored observations that have not been considered in the analysis. It is essential to consider the censored observations for predicting the lifetime of the component. In this situation, the expectation maximization (EM) algorithm can be applied to obtain the ML estimates of the parameters of the model because of incomplete information on covariates (Karim and Suzuki 2007). Finally, the database of the automobile component contains information on manufacturing month. An important and useful extension of the present research is to employ the D&R method by considering divisions over batches (manufacturing months) and compare the results.

Table 4 MLEs of parameters from simulated data

Coefficients/Parameters	All data 100,000 observation	200 observations	400 observations	600 observations	800 observations
Intercept (β_0)	9.652065	9.639672	9.647097	9.646629	9.653301
Age (β_1)	0.001997	0.001983	0.001986	0.001997	0.001979
Region2 (β_2)	−0.230666	−0.227143	-0.2308325	−0.230348	−0.232019
Region3 (β_3)	−0.241675	−0.234832	−0.2414701	−0.245749	−0.239832
Region4 (β_4)	−0.161844	−0.156937	−0.1622794	−0.159916	−0.161840
FM2 (β_5)	0.029894	0.0257678	0.0306917	0.028722	0.029499
FM3 (β_6)	0.004708	0.0048184	0.0015547	0.004803	0.003271
Percentile					
10th percentile	619.41	723.46	667.76	652.38	642.50
25th percentile	3615.46	3973.98	3783.45	3732.12	3695.85
50th percentile	16,941.27	17,789.04	17,338.17	17,226.66	17,130.99
75th percentile	57,236.41	58,260.40	57,702.14	57,611.34	57,457.30

Source Authors' own calculation by using R

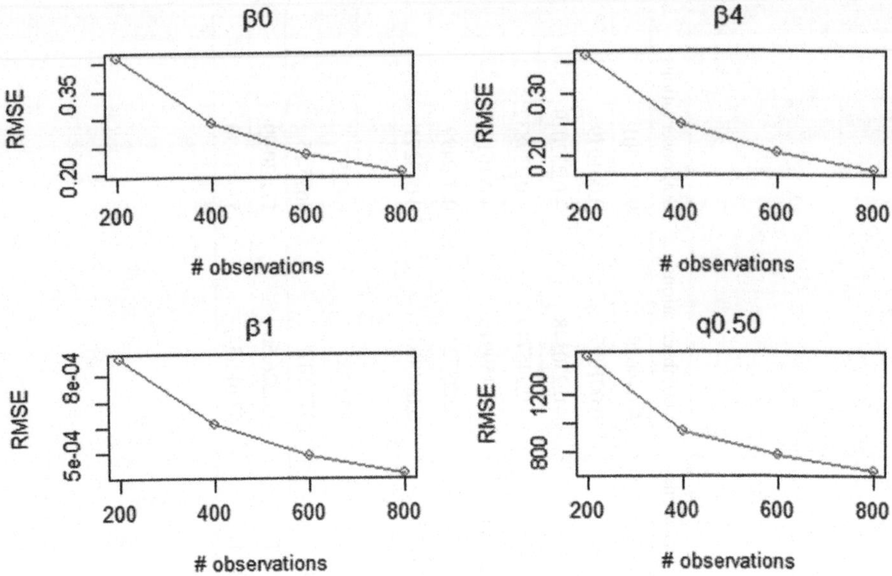

Fig. 2 RMSE of the MLEs from simulated data. *Source* Created by the authors

Acknowledgements The authors are grateful to the reviewer for his comments and suggestions on an earlier version of the manuscript.

Appendix: Programming Codes in R

This appendix provides the programming codes in R (Web site http://cran.r-project.org/) that are applied to analyze the data in the example discussed in Sect. 4. A portion of the data frame, for example, given in Table 5 is used in the program.

In this example, the data frame named df consists of the following variables:

Table 5 A portion of the data frame (automobile data) that is used in the example

Mileage	Age	Region1	Region2	Region3	Region4	Fm1	Fm2	Fm3
48,023	479	0	0	0	1	0	1	0
28,931	506	1	0	0	0	0	1	0
21,308	368	0	1	0	0	0	1	0
22,050	463	0	0	1	0	0	1	0
15,295	193	0	0	0	1	1	0	0
19,285	293	1	0	0	0	0	0	1

Source Authors' own processing

- Mileage: usage km, the dependent variable in the regression model,
- Age: age in days,
- Region1, Region2, Region3, Region4: automobile used regions (four different regions, dummy variables), and
- Fm1, Fm2, Fm3: failure mode (three different failure modes, dummy variables).

```
library(dummies)
library(survival)
library(SPREDA)

# fit.comp: function for estimating parameters using all
# data at a time
# x: vector of covariate values; p: vector of
# probabilities for quantiles.
fit.comp <- function(df,x,p){
  fit01 <- survreg(Surv(df[,1]) ~ df[,2] + df[,4] +
  df[,5] + df[,6] + df[,8] + df[,9], dist="weibull")
  b <- fit01$coef; sigma <- fit01$scale
  se <- summary(fit01)$table[,2]; se<-round(se, digits=6)
  conf.intvl <- confint(fit01, level=0.95)
  mu.hat <- t(b)%*%x
  mu.hat.vec <- rep(mu.hat, length(p))
  tp <- exp(mu.hat.vec+qsev(p)*sigma)
  estimate <- list(Coefficients=b, Std.Errors=se,
  Conf.intervals=conf.intvl, Quantiles=tp)
  return(estimate)
 }
```

```
# fit.DR: function for estimating parameters through
# D&R method with rep.time random divisions
# n.coef: No. of regression coefficients in the model
fit.DR <- function(df,rep.time,x,p){
  n.coef <- 7  # No. of regression coeff. in the model
  b <- matrix(NA,rep.time,n.coef)
  sigma <- array(NA,rep.time)
  se <- matrix(NA,rep.time,n.coef)
  tp <- matrix(NA,rep.time,length(p))
  sum.conf.intvl <- 0
  for(i in 1:rep.time){
    df.sam <- df[sample(nrow(df), 500, replace = TRUE), ]
    fit.sam <- survreg(Surv(df.sam[,1]) ~ df.sam[,2] +
     df.sam[,4] + df.sam[,5] +
    df.sam[,6] + df.sam[,8] + df.sam[,9], dist="weibull",
    data = df.sam)
    b[i,] <- fit.sam$coef; sigma[i] <- fit.sam$scale
    se[i,] <- summary(fit.sam)$table[,2][-(n.coef+1)]
    CI <- confint(fit.sam, level=0.95)  # 95% conf. intv.
    sum.conf.intvl <- sum.conf.intvl + CI
    mu.hat <- t(b[i,])%*%x
    mu.hat.vec <- rep(mu.hat, length(p))
    tp[i,] <- exp(mu.hat.vec+qsev(p)*sigma[i])
  }
  Beta <- colMeans(b)
  avg.se <- colMeans(se)
  avg.CI <- sum.conf.intvl/rep.time
  tp.hat <- colMeans(tp)
  est.DR<-list(Coefficients=Beta, Std.Errors=avg.se,
  Conf.intervals=avg.CI,
  Quantiles=tp.hat)
  return(est.DR)
}
```

References

Blischke, W. R., Karim, M. R., & Murthy, D. N. P. (2011). *Warranty data collection and analysis*. Springer-Verlag.

Cleveland, S., & Hafen, R. (2014). Divide and Recombine (D&R): Data science for large complex data. *Statistical Analysis and Data Mining, 7*, 425–433.

Guha, S., Hafen, R., Rounds, J., Xia, J., Li, J., Xi, B., & Cleveland, W. (2012). Large complex data: Divide and Recombine (D&R) with rhipe. *Stat, 1*(1), 53–67.

Hafen, R. (2016). Divide and recombine: Approach for detailed analysis and visualiza-tion of large complex data. In P. Bühlmann, P. Drineas, M. Kane, & M. V. Laan (Eds.), *Handbook of big data* (pp. 35–46). Chapman & Hall, CRC.

Karim, M. R., & Islam, M. A. (2019). *Reliability and survival analysis*. Springer Nature Singapore Pte Ltd.

Karim, M. R., & Suzuki, K. (2006). Analysis of warranty data with covariates. In W. Y. Yun & T. Dohi (Eds.), *Advanced reliability modeling II—reliability testing and improvement* (pp. 377–384). World Scientific Publishing Co.

Karim, M. R., & Suzuki, K. (2007). Analysis of warranty data with covariates. *Proceedings of the Institute of Mechanical Engineering, Part o, Journal of Risk and Reliability, 221*(4), 249–255.

Lawless, J. F. (2003). *Statistical models and methods for lifetime data* (2nd ed.). John Wiley & Sons Inc.

Lee, J. Y., Brown, J. J., & Ryan, M. M. (2017). Sufficiency revisited: Rethinking statistical algorithms in the big data era. *The American Statistician, 71*(3), 202–208.

Liu, W., & Li, Y. (2018). A new stochastic restricted Liu estimator for the logistic regression model. *Open Journal of Statistics, 8*, 25–37.

Meeker, W. Q., & Escobar, L. A. (1998). *Statistical methods for reliability data.* Wiley Interscience.

Performance of Different Data Mining Methods for Predicting Rainfall of Rajshahi District, Bangladesh

Md. Mostafizur Rahman, Md. Abdul Khalek, and M. Sayedur Rahman

Abstract Rainfall predicting by efficient method is always interesting for particular region because timely and accurately forecasted rainfall data is extremely helpful to take necessary safety action in advance, in case of agricultural production, flood management, drought monitoring, and ongoing construction project. Data mining technique is suitable for predicting different environmental attributes by extracting new relationships from the past data. So, researchers are always trying to predict rainfall data with maximum accuracy by optimizing and integrating different data mining techniques for different weather stations. In our study, we compare the forecasting performance of Linear Discriminant Analysis, Classification and Regression Trees, Random Forest, K-Nearest Neighbors, and Support Vector Machine for rainfall prediction, in case of Rajshahi district, Bangladesh. The monthly time series data for the time period January, 1964 to December, 2017 is considered for analysis. Data mining processes such as data collection, data pre-processing, modeling, and evaluation are strictly followed for empirical studies. The forecasting performances of these models are confirmed by precision, recall, f-measure, and overall accuracy, and also by graphical method. The empirical result showed that the k-nn method is the most suitable method for predicting rainfall in case of Rajshahi district, Bangladesh for the subsequent time period.

Keywords Linear discriminant analysis · Classification and regression trees · Random forest · k-nearest neighbor · Support vector machine · Rainfall

Md. M. Rahman (✉) · Md. A. Khalek · M. S. Rahman
Environment and Data Mining Research Group, Department of Statistics, University of Rajshahi, Rajshahi 6205, Bangladesh

B. K. Sinha et al. (eds.), *Data Science and SDGs*,
https://doi.org/10.1007/978-981-16-1919-9_6

1 Introduction

In time series analysis, application of data mining technique is quite interesting and recently gained popularity. In such case, data is collected either in hourly, daily, weekly, monthly, quarterly, and yearly or other specified time periods (Mishra et al. 2018). Data mining technique tried to discover new relationships from different attributes in particular cases and it is useful for analyzing financial data, environmental data, time series data, and so on (Gupta and Ghose 2015; Ahmad et al. 2017).

Rainfall is a natural phenomenon which collects droplets from atmospheric water vapor, becomes heavy, and finally falls on the ground. It is crucial for lives and agricultural production. Timely and perfect prediction of rainfall data alert the policy maker to take necessary steps in advance in case of agricultural production, flood control, water management, flight operation, and upcoming construction works (Chau and Wu 2010; Wu et al. 2015). In developing predictive model for rainfall data, researchers face problem for determining the contribution of atmospheric process due to the uncertainty. Humidity, maximum and minimum temperature, wind speed and direction, and cloud coverage are the important influencing factors for rainfall. In case of predicting rainfall efficiently, many researchers proposed different data mining techniques; for example, Olaiya and Adeyemo (2012) investigated the performance of artificial neural network and decision tree algorithm for predicting rainfall, maximum temperature, wind speed, and evaporation in case of Nigeria and found good performance of artificial neural network over decision tree algorithm for all of these attributes. Ramana et al. (2013) compared the performance of wavelet neural network (WNN) model, which is an integration of wavelet technique and artificial neural network, with traditional artificial neural network (ANN) model for predicting rainfall in case of Darjeeling of India and found that their proposed WNN model perform better than traditional ANN model. Zainudin et al. (2016) investigated the performance of Naïve Bayes, Decision Tree, Random Forest, Support Vector Machine, and Neural Network model for predicting Malaysian rainfall data and found that Random forest model provided the most successful result than other methods for predicting rainfall data in case of Malaysia. Bagirov et al. (2017) proposed Cluster base Linear Regression (CLR) method for predicting rainfall data of eight weather stations in Australia and compared the forecasting performance of their proposed model with artificial neural network, multiple linear regression, and support vector machine. Their analytical result showed that CLR model gives better forecasting result than other models in most of the locations. Cramer et al. (2017) compared the performance of Markov Chain model with some machine learning techniques for predicting the rainfall data of 42 cities and concluded that all of the machine algorithms perform better than existing Markov Chain model. Solanki and Panchal (2018) proposed hybrid Artificial Neural network model from the combination of Artificial Neural Network and Genetic Algorithm and found that their proposed model showed better forecasting performance than other existing models. Mishra et al. (2018) developed one-month and two-month ahead forecasting models by using ANN model in case of different weather stations of North India and found that one-month ahead forecasting model gave better forecasting than two-month ahead forecasting model. Aftab et al. (2018a) examined the performance of Naïve Bayes (NB), Support Vector Machine (SVM), Decision Tree, k-Nearest Neighbor (k-NN), and Multilayer Perceptron (MLP) algorithm for predicting rainfall data in case of Lahore city and found that these techniques performed well for no-rain class, but in case of rain class, it fails to show better performance. Beside these, different authors such as Sivapragasam et al. (2001); Monira et al. (2010); Kannan et al. (2010); Sethi and Garg (2014); Talib et al. (2017); Tharun et al. (2018); Aftab et al. (2018b) investigated the performance of different data mining models and techniques for forecasting rainfall data for specific cities or regions.

From the above discussion, we found that researchers are consistently working hard to find out the most successful predictive model by optimizing and integrating different data mining techniques in different weather stations, and found different models showing better performance for different weather stations. Predicting rainfall by different data mining techniques for Rajshahi District is rare. So, in this paper, we compare the forecasting performance by different data mining techniques in case of Rajshahi District, Bangladesh. This study will be helpful to give prior information about rain, flood, and drought, which will save lives and properties of the people and contribute for the development of the economy. The paper is organized as follows: Sect. 1 gives the introduction, Sect. 2 presents methods and materials, Sect. 3 gives result and discussion, and finally, Sect. 4 concludes the paper.

2 Materials and Methods

Rajshahi is situated in the north-west part of Bangladesh covering an area around 2407.01 sq km, its location is from 24°07′ to 24°43′ north latitudes and from 88°17′ to 88°58′ east longitudes. It is surrounded by Chapai Nawabganj district on the west, Natore district on the east, Naogaon district on the north, and Kushtia district on the south. The climate of the study area is classified as tropical. Usually maximum amount of rainfall is observed in summer. The amount of yearly rainfall is about 1419 mm. The maximum and minimum temperature of this area is about 29.4 °C and 18.5 °C respectively. It consists of barind, diara, and char type of lands. A classification framework for data mining process is given in Fig. 1.

Figure 1 presents the main steps of data mining techniques: Data collection where data was collected from data sources, Data Pre-processing which involve data cleaning and data transformation, Prediction where different data mining techniques were applied for perdiction, and Model evaluation which compares the performance of predictive models.

Fig. 1 Data mining process (*Source* created by authors)

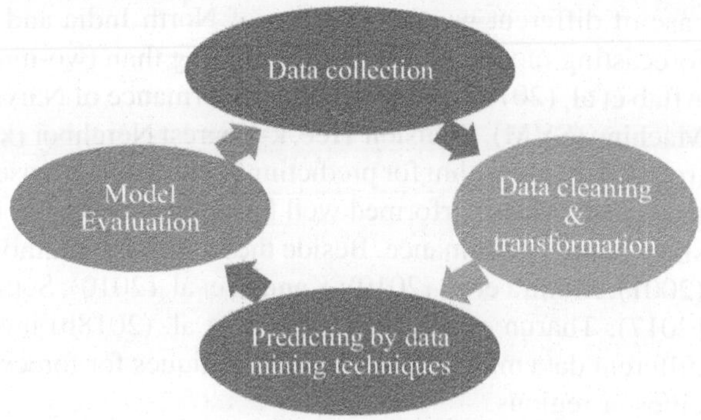

Table 1 Nature of Data [attributes with their measurements/scales]

Variables	Type	Measurement	Missing values
Rainfall status	Categorical	(yes/no)	00
Temperature	Numerical	Degrees celsius	09
Humidity	Numerical	%	32
Wind speed	Numerical	Meters/second	15
Sunshine	Numerical	Hour	10
Max.Temp	Numerical	Degrees celsius	24
Min.Temp	Numerical	Degrees celsius	20

Source Authors' own compilation

2.1 Data Collection

The data were collected from Bangladesh Meteorological Department (BMD), and it covered monthly data for the time period from January, 1964 to December, 2017. The input dataset obtained for rainfall prediction consist of several atmospheric attributes such as temperature, humidity, wind speed, sunshine, minimum temperature and maximum temperature. See Table 1 below.

2.2 Data Pre-Processing

The performance of analytical result depends on the quality of data. Data pre-processing stage ensures the quality of data. This stage performs the following four activities: Data cleaning, Data transformation, Data integration, and Data reduction. The data from Table 1 indicate that all of the variables [except rainfall status] are quantitative and have different measurement scales, and there are also some missing values for each of them. The dataset also contain some noisy data where the values lie lower a certain limit or exceed a certain limit. The missing value has been filled up by the smoothing technique. Since our data sets contain different measuring scales, so we transform all of the variables by Minimum Maximum Normalization method, which confirms the entire set of variables converted to the same range.

2.3 *Predictive Models*

To predict rainfall data accurately, we need to investigate the performance of some predictive models. These are described below.

2.3.1 Classification and Regression Trees (CART)

The CART algorithm was introduced by Breiman et al. in 1984 where the classifiers are strictly binary, containing exactly two branches for each decision node. It partitions the record of the training set into subsets of records with similar values, in case of dependent or target variable. The decision tree grows up by conducting for each decision node for all possible available variables and splitting value. The optimal split is selected by the following criteria. Let $\Phi(s|t)$ be the measure of the "goodness" of a candidate split and defined as:

$$\Phi(s|t) = 2P_L P_R \sum_{j=1}^{\#\text{classes}} |P(j|t_L) - P(j|t_R)|$$

where t_L and t_R are the left and right child node respectively at node t,
P_L =(number of records at t_L)/(number of records in training set),
P_R =(number of records at t_R)/(number of records in training set),
$P(j|t_L)$ =(number of class j records at t_L)/(number of records at t),
$P(j|t_R)$ = (number of class j records at t_R)/(number of records at t).

Then the initial root is chosen from the split where the $\Phi(s|t)$ produce the maximum value.

2.3.2 Support Vector Machine (SVM)

Support Vector Machines are useful tools for classification and prediction for both linear and non-linear cases. It uses non-linear mapping to transform the original training data into a higher dimension. It looks for linear optimal separating hyperplane within this new dimension. In case of two linear separable classes, let $\{(X_1, y_1), (X_2, y_2), \ldots, (X_{|D|}, y_{|D|})\}$, where D is the data set and X_i is the set of training tuples with associated class labels y_i. Each class label can take either +1 or −1 (i.e., $(y_i \epsilon \{+1, -1\})$, corresponding to the class, yes and no, respectively. Let us

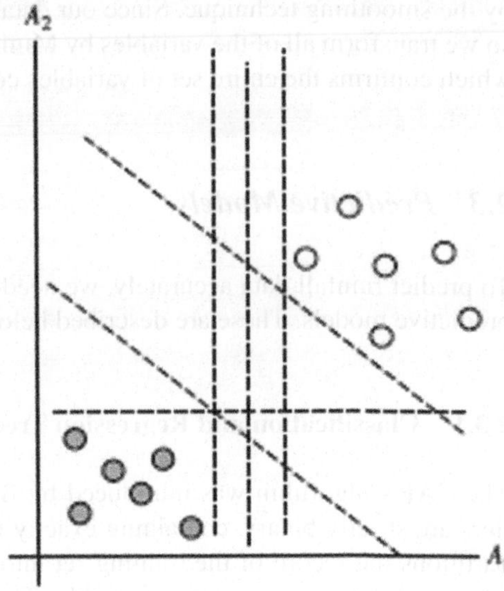

Fig. 2 Diagram for support vector machine (*Source* Han and Kamber 2006)

consider two input attributes, X_1 and X_2, where class +1 is indicated by light circle and class -1 by gray circle (Fig. 2).

The above figure shows the linearly separable case because straight can drawn to separate these two classes. It is possible to draw infinite number of straight lines to separate these two classes. We need to choose the best class boundary, which classifies the classes properly with minimum classification error.

2.3.3 K-Nearest Neighbors (K-NN)

Although the K-nearest neighbor algorithm was introduced in the early 1950, but until 1960 it did not receive popularity due to labor intensity. It works by comparing a given test data with training data, which are close or similar. The training data are described by the total number of n attributes where each data represent a point in n-dimensional space. By continuing this process, all of the training data are stored in n-dimensional space. When any new observation comes, then K-nn algorithm searches the pattern which are more nearest to the new observation. This similarity is measured by Euclidean distance.

Let $X_1 = (x_{11}, x_{12}, \ldots, x_{1n})$ and $X_2 = (x_{21}, x_{22}, \ldots, x_{2n})$ be two data points, then the distance can be calculated as: $distance(X_1, X_2) = \sqrt{\sum_{i=1}^{n} (x_{1i} - x_{2i})^2}$. This similarity is considered which give smaller distance. Before measuring the distance, the data need to be converted into normalization (Han and Kamber 2006).

2.3.4 Random Forest

In classification purposes, random forest algorithms are widely used. The random forest algorithm works in the concept of boosting techniques. This algorithm first creates some number of randomly generated decision tree on subsets of the data; after that it uses averaging technique for improving the classification accuracy and reduced the over fitting problem. The trees are grown up by splitting the attributes set into its subset, and it is continuing until a subset in a given node is equal to the response variable. In each stage, the average fit into new learner to the difference observed, and at every step, the ensemble fits a new learner to the difference between the observed and total prediction. In every fit, the mean squared error is minimized. The prediction is made by taking the average prediction from all learners.

2.3.5 Linear Discriminant Analysis (LDA)

The Linear Discriminant Analysis is a very simple model, both in preparation and application. The mean and standard deviation is calculated for each single variable (x) for each class. The mean and the covariance matrix is calculated over the multivariate Gaussian in case of multiple variables. From the data, all of the statistical properties are estimated and then plug it into the linear discriminant analysis for prediction. LDA makes some simplifying assumptions:

i) The data is to be Gaussian, i.e., each variable shows a bell-shaped curve.
ii) All of the attributes present the same variance.

With these assumptions, the mean and variance at every class of data is estimated by the LDA model. It makes prediction by estimating the probability of a new set of input variables from each class. The prediction is complete when the class receive the highest probability in output class.

2.4 *Model Evaluation*

The precision, recall, and F-measure are well-known statistics for comparing the forecasting performance of different predictive models. Overall accuracy, which is calculated as the proportion of the total number of prediction, is also used here.

Precision

The "Precision" evaluates the True Positive (TP) entities with respect to the False Positive (FP) entities. The estimated formula for precision is given below:

$$\text{precision} = \frac{TP}{(TP + FP)}$$

Here, TP indicates correctly classified and FP indicates wrongly classified.

Recall

The "Recall" evaluates the True Positive (TP) entities with respect to the False Negative (FN) entities. This can be estimated by the following formula: Recall $= \frac{TP}{(TP+FN)}$

F-Measure

If any model has high precision value and low recall value, then the questions arise which model is better. In this case, precision and recall value cannot uniquely identify the model. For solving this kind of problem, we use F-measure which is defined as follows: $F - \text{measure} = \frac{2(\text{Precision})(\text{Recall})}{(\text{Precision}+\text{Recall})}$

3 Results and Discussions

In order to find out the most suitable predictive model, first, we perform data pre-processing phase, and after that we compared the forecasting performance of different data mining algorithms, such as Linear Discriminant Analysis (LDA), Classification and Regression Trees (CART), Random Forest (RF), K-Nearest Neighbors (K-NN), and Support Vector Machine (SVM). All of these predictive models are treated as supervised methods. Performance of any supervised method is analyzed by comparing the output result with pre-classified data. In our study, we consider the percentage of training and test data are 70% and 30%, respectively. The summaries for the accuracy of different methods are given in Table 2. The mean statistics for K-NN method showed higher performance compared to other models for predicting the rainfall data of Rajshahi, Bangladesh, whereas Random Forest models showed the worst performance.

In model evaluation phase, different measures such as precision, recall, f-measure, and overall accuracy are applied for checking the forecasting performance of these models. Table 3 reported the calculated precision value, recall value, and f-measure value for both rain and no-rain class, and also overall classification accuracy for all these models. The estimated results from Table 3 confirmed that the K-NN method

Table 2 Summary for accuracy measurement

Model	Min	1st Quar.	Median	Mean	3rd Quar.	Max
CART	0.7731959	0.7938144	0.7938144	0.8024195	0.8247423	0.8265306
SVM	0.7938144	0.8041237	0.8061224	0.8148327	0.8247423	0.8453608
K-NN	0.7628866	0.7857143	0.8125000	0.8168025	0.8453608	0.8775510
RF	0.7525773	0.7938144	0.7959184	0.7983589	0.8144330	0.8350515
LDA	0.7525773	0.7731959	0.7959184	0.8066064	0.8453608	0.8659794

Source Authors' own calculation

Table 3 Model evaluation criteria

Model	Class	Precision	Recall	F-measure	Overall accuracy
CART	Yes	0.8889	0.9285	0.9082	0.9130
	No	0.9167	0.8594	0.8871	
SVM	Yes	0.9091	0.9278	0.9183	0.9006
	No	0.8871	0.8594	0.8730	
K-NN	Yes	0.9200	0.9478	0.9336	0.9144
	No	0.9190	0.8838	0.9010	
RF	Yes	0.9091	0.9278	0.9183	0.9006
	No	0.8871	0.8594	0.8730	
LDA	Yes	0.8812	0.9175	0.8989	0.8758
	No	0.8667	0.8125	0.8387	

Source Authors' own calculation

showed better forecasting performance for predicting rainfall data in case of Rajshahi, Bangladesh in case of both "yes" and "no" classes. Based on overall accuracy, these predictive models may be ranked as follows: K-NN > CART > RF > SVM > LDA, where RF and SVM model showed similar overall accuracy. Besides these model evaluation criteria, we also consider other model evaluation criteria based on graphical measurement such as Box and Whisker Plot, density plot, dot plot, and parallel plot which is shown in Fig. 3.

From Fig. 3 we may conclude the following:

(1) The graphical view from Box and Whisker plots of Fig. 3 presents the order from highest to lowest mean accuracy. We found that the overall accuracy of K-NN method is higher compared to other models.

(2) The density plots showed the distribution of the model accuracy, and it is useful for evaluating the overlap in the estimated behavior of algorithms. We like to look at the differences in the peaks as well as the spread or base of the distributions. The density plot for K-NN method presents the highest peak.

(3) These plots presented the mean estimated accuracy and 95% confidence interval. It is necessary to compare the means and eye-ball the overlap of the spreads between algorithms. Finally, parallel plots also confirm better performance of K-NN method than other methods.

4 Conclusions

Rainfall has a great impact on agriculture and economy all over the world. Data mining techniques are successful for predicting rainfall data by extracting new relationships from the existing data. Nowadays, developed models and techniques are available for predicting rainfall data but still there are no unique models or techniques

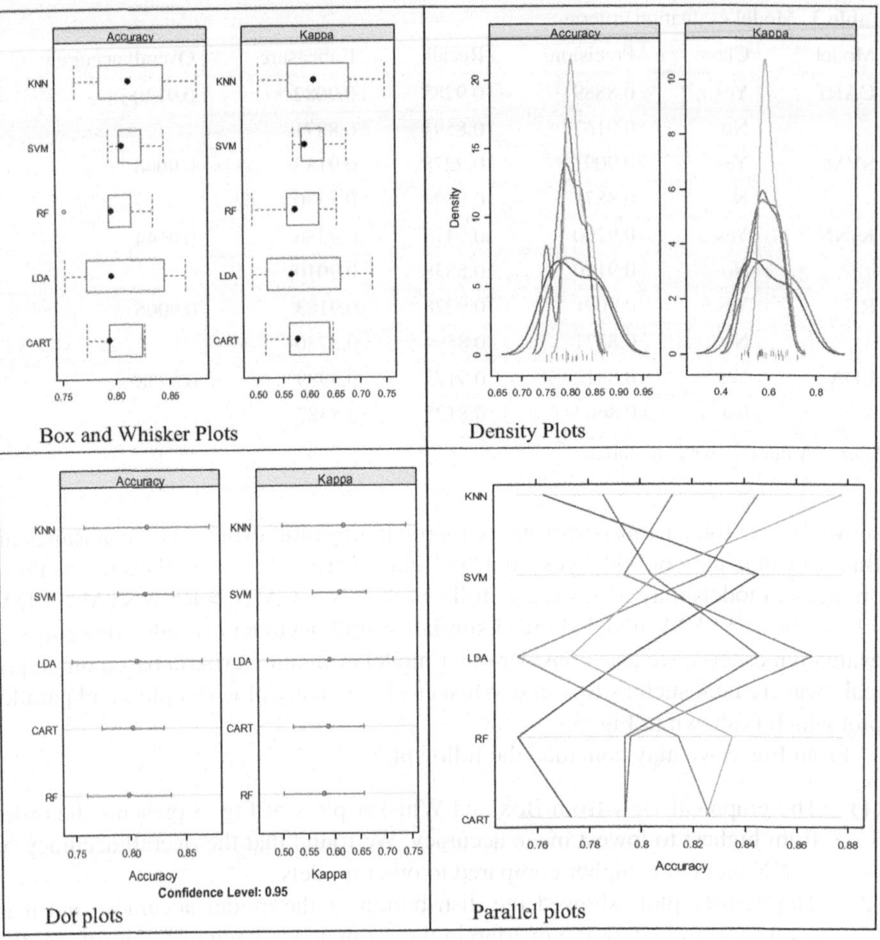

Fig. 3 Plot for measuring accuracy (*Source* Created by authors)

which can predict the rainfall data accurately for all the geographical locations. So, in this paper, we compared the forecasting performance of different data mining techniques, such as Linear Discriminant Analysis, Classification and Regression Trees, Random Forest, K-Nearest Neighbors, and Support Vector Machine, for rainfall prediction in Rajshahi District, Bangladesh for the time period January, 1964 to December, 2017 on a monthly basis. To predict properly and accurately, data pre-processing methods are applied which perform data cleaning, filling missing value, transforming, etc. The predicting ability of these models is examined by some standard measures for comparing forecasting performance, and it is also verified by different graphical comparison tools. The empirical results confirmed that the K-Nearest Neighbor method showed the highest performance for predicting the rainfall data for this time period, whereas Linear Discriminant Analysis method presented a poor predicting performance. So, we may conclude that the K-NN method is the most suitable method for predicting rainfall data in the case of Rajshahi District, Bangladesh for the subsequent time periods. The findings in this paper are not unique, it may vary for different time periods and different geographical locations. This is for future work. The findings will help policy makers or researchers to take necessary steps in advance to face the upcoming situation.

Acknowledgements The authors highly appreciate the constructive suggestions and comments from distinguished reviewers in the earlier version of this paper.

References

Aftab, S., Ahmad, M., Hameed, N., Bashir, M. S., Ali, I., & Nawaz, Z. (2018a). Rainfall prediction in Lahore City using data mining techniques. *International Journal of Advanced Computer Science and Applications, 9*(4), 254–260.

Aftab, S., Ahmad, M., Hameed, N., Bashir, M. S., Ali, I., & Nawaz, Z. (2018b). Rainfall prediction using data mining techniques: A systematic literature review. *International Journal of Advanced Computer Science and Applications, 9*(5), 143–150.

Ahmad, M., Aftab, S., & Muhammad, S. S. (2017). Machine learning techniques for sentiment analysis: A review. *International Journal of Multidisciplinary Sciences and Engineering, 8*(3), 27–32.

Bagirov, A. M., Mahmood, A., & Barton, A. (2017). Prediction of monthly rainfall in Victoria, Australia: Cluster wise linear regression approach. *Atmospheric Research, 188*, 20–29.

Breiman, L., Friedman, J., Olshen, R., & Stone, C. (1984). *Classification and regression trees* (p. 1984). Chapman & Hall/CRC Press.

Chau, K. W., & Wu, C. L. (2010). A hybrid model coupled with singular spectrum analysis for daily rainfall prediction. *Journal of Hydroinformatics, 12*(4), 458.

Cramer, S., Kampouridis, M., Freitas, A. A., & Alexandridis, A. K. (2017). An extensive evaluation of seven machine learning methods for rainfall prediction in weather derivatives. *Expert Systems with Applications, 85*, 169–181.

Darji, M. P., Dabhi, V. K., & Prajapati, H. B. (2015). Rainfall forecasting using neural network: A survey. In *International Conference on Advances in Computer Engineering and Applications* (pp. 706–713).

Gupta, D., & Ghose, U. (2015). A comparative study of classification algorithms for forecasting rainfall, 2015 (pp. 0–5).

Han, J., & Kamber, M. (2006). *Data mining concepts and techniques* (2nd ed.). Morgan Kaufmann Publishers.

Kannan, M., Prabhakaran, S., & Ramachandran, P. (2010). Rainfall forecasting using data mining technique. *International Journal of Engnieering and Technology, 2*(6), 397–401.

Larose, D. T. (2006). *Data mining methods and models*. John Wiley & Sons.

Mishra, N., Soni, H. K., Sharma, S., & Upadhyay, A. K. (2018). Development and analysis of artificial neural network models for rainfall prediction by using time-series data. *International Journal of Intelligent Systems and Applications, 10*(1), 16–23.

Monira, S. S., Faisal, Z. M., & Hirose, H. (2010). Comparison of artificially intelligent methods in short term rainfall forecast. In 2010 13th International Conference on Computer and Information Technology, ICCIT 2010 (pp. 39–44).

Olaiya, F., & Adeyemo, A. B. (2012). Application of data mining techniques in weather prediction and climate change studies. *International Journal of Information Engineering and Electronic Business*, 1, 51–59.

Ramana, R. V., Krishna, B., Kumar, S. R., & Pandey, N. G. (2013). Monthly rainfall prediction using wavelet neural network analysis. *Water Resource Management, 27*(10), 3697–3711.

Sivapragasam, C., Liong, S., & Pasha, M. (2001). Rainfall and runoff forecasting with SSA-SVM approach. *Journal of Hydroinformatics*, April 2016, 141–152.

Sethi, N., & Garg, D. K. (2014). Exploiting data mining technique for rainfall prediction. *International Journal of Computer Science and Information Technologies, 5*(3), 3982–3984.

Solanki, N., & Panchal, G. (2018). A Novel Machine Learning Based Approach for Rainfall Prediction. In *International Conference on Information and Communication Technology for Intelligent Systems (ICTIS 2017)*, vol. 1, vol. 83.

Talib, M. R., Ullah, T., Sarwar, M. U., Hanif, M. K., & Ayub, N. (2017). Application of data mining techniques in weather data analysis. *International Journal of Computer Science and Network Security, 17*(6), 22–28.

Tharun, V. P., Prakash, P., & Devi, S. R. (2018). Prediction of rainfall using data mining techniques. In *Proceeding of the 2nd International Conference on Inventive Communication and Computational Technologies (ICICCT 2018) IEEE Explore Compliant-Part Number: CFP18BAC-ART*, ISBN 978-1-5386-1974-2.

Vathsala, H., & Koolagudi, S. G. (2017). Prediction model for peninsular Indian summer monsoon rainfall using data mining and statistical approaches. *Computers & Geosciences, 98*, 55–63.

Wu, J., Long, J., & Liu, M. (2015). Evolving RBF neural networks for rainfall prediction using hybrid particle swarm optimization and genetic algorithm. *Neurocomputing, 148*, 136–142.

Zainudin, S., Jasim, D. S., & Bakar, A. A. (2016). Comparative analysis of data mining techniques for Malaysian rainfall prediction. *International Journal on Advanced Science Engineering and Information Technology, 6*(6), 1148–1153.

Generalized Vector Autoregression Controlling Intervention and Volatility for Climatic Variables

Md. Ashek Al Naim, Md. Abeed Hossain Chowdhury, Md. Abdul Khalek, and Md. Ayub Ali

Abstract The purpose of this study is to build a time series model for forecasting the climatic variables of Rajshahi district using the VAR model controlling intervention and volatility. Seven models for seven climatic variables are found, and the stability of every model is checked with proper validation techniques. The fitted models are GVAR with GARCH (2,1) and intervention for Cloud coverage; GVAR with GARCH (3,1) and intervention for Relative Humidity; ARIMA (1,0,1) with GARCH (1,1) for rainfall, GVAR with GARCH (2,1), and intervention for maximum Temperature; GVAR with ARCH (2) and intervention for minimum temperature; GVAR with intervention for sunshine; and ARIMA (2,0,2) for wind speed. The stable models are used to forecast the daily data which may be beneficial to people and policymakers. Finally, it is found by forecasting that Maximum Temperature (T1), Humidity (H), Bright Sunshine (S), and Wind Speed (W) might be shown upward trend while Minimum Temperature (T2), Rainfall (R), and Cloud Coverage (Cl) might be shown decreasing trend from the year 2018 to 2022. Considering the finding of this study, Government and policymakers can make people aware of the adverse effect of climate change.

Keywords VAR · ARCH · GARCH · ARIMA · GVAR · Stationarity · Stability · Climatic variables · Climate change

1 Introduction

Climate change refers to the long-term variation in weather studied by many researchers around the world (Shahin et al. 2014; Ferdous et al. 2014; Khan and Ali 2003; Kleiber et al. 2013). Their studies lead them to forecast upcoming climate changes. Forecasting is really an interesting thing, and it helps us to understand the situation in advance. It is very important in the sense that it may save many lives

Md. A. A. Naim · Md. A. Khalek · Md. A. Ali (✉)
Department of Statistics, University of Rajshahi, Rajshahi 6205, Bangladesh

Md. A. H. Chowdhury
BARC, Dhaka, Bangladesh

B. K. Sinha et al. (eds.), *Data Science and SDGs*,
https://doi.org/10.1007/978-981-16-1919-9_7

and belongings from natural disasters like flood, drought, northwester, hurricane, cyclone, heavy rainfall or no rainfall, etc.

Bangladesh is a land where natural disaster is a common matter. Rajshahi district of Bangladesh is an extreme climatic district. High temperature in summer, low temperature in winter, less rainfall, high humidity, etc. are the climatic properties of Rajshahi. Every year during the rainy season, the people of Bangladesh suffer from the agony of floods. The northwester is also a common problem resulting in cyclones like Sidr in 2007, Reshmi in 2008, Bijli in 2009, Aila in 2009, and Fani in 2019 that have cost many valuable lives and belongings of people especially people in the coastal area of Bangladesh. The loss of life and wealth is enumerated by many researchers (Cutter and Finch 2008; Dilley et al. 2005; and Hallegatte and Przyluski 2010).

The farmers as well as others will be highly benefitted if they come to know the weather report beforehand.

There are lot of works already done on climate change. Ahmed and Hossen (2018) studied the temporal and spatial variation of the absolute maximum temperature of Bangladesh. Ferdous et al. (2014) built a time series model and forecast on the maximum temperature data. Shamsina and Hossain (2011) used the stochastic method for modeling weather patterns. Kleiber et al. (2013) studied the climate impact of minimum and maximum temperature using spatiotemporal simulation. Khan and Ali (2003) proposed a method of VAR modeling when the variables are mixed in nature. Shahin et al. (2014) used the VAR model for forecasting temperature, humidity, and cloud coverage on monthly basis data. Considering all these literature studies, we see no work with every climate variable on daily basis data.

So, a study on every climatic variable using daily basis data is needed. From the study, we may be able to get new clues that may help us to go forward and to explore the interrelation among all the climatic variables. The information gained from the study may help to raise awareness before anything unexpected happens.

An attempt is taken in this study for exploring such a time series model for furcating the climatic variables of Rajshahi district using the VAR model.

2 Materials

The daily data of climate variables maximum temperature (T1), minimum temperature (T2), wind speed (W), Relative Humidity (H), Cloud Coverage (CL), Bright Sunshine (S), and Rainfall (R) from the year 2000 to 2017 are taken from "Database and Statistics, Bangladesh Agricultural Research Council" BARC (www.barc.gov.bd/data_stat.php). Of a total of 6575 observations, only 5844 observations are used to develop the VAR models and the rest 731 observations are used to validate the models. For the computational work, software like MS Excel, E-views, and R is used. MS word is used to store the findings in black and white.

3 Methods

We want to forecast all the climatic variables noted in the materials section. As the variables are intercorrelated, so a vector autoregression (VAR) model is applicable. For building a VAR model, the first step is to check the stationarity of these variables. Here to test, the stationarity graphical representation and correlograms are used to see if there is any trend in the series. The popular unit root test called ADF test (Dickey and Fuller 1979), PP test (Phillips and Perron 1988), and KPSS test (Kwiatkowski et al. 1992) is also performed to ensure the unit root in these series (Gujarati 1995). Endogeneity of variables needed to confirm a VAR model in which all the variables should be endogenous. According to Gujarati (1993), all the variables under the study in a VAR model are endogenous and usually, no one is exogenous. For sorting out the endogenous variables, a pairwise Granger Causality test is performed. To build the VAR(P) model, lag P determination is important. The order of lag is usually determined by Akaike Information criteria (AIC) in 1974, Schwarz Criteria (SC) in 1978, Hannan Quinn information criteria (HQ) in 1979, Final prediction error (FPE) (Akaike, 1969), and Likelihood ratio statistics (Lütkepohl 1991). For estimating VAR(P) model, the usual OLS method can be applied to each equation. The OLS method needed to apply to k equations to estimate parameters. After the model being identified and its parameter estimated, only the significant parameters are kept. The heteroscedasticity of the models is checked. Such a model incorporating the possibility of a non-constant error variance is called a heteroscedastic model (Wei 2006). Diagnostic checking of each of the models is performed. Time series can be affected by an extreme value, which is why intervention analysis is also performed where necessary. It is extremely sensitive to the violation of the independence assumption as shown by Box and Tiao (1975), who developed the intervention analysis (Wei 1990).

4 Results

At first, the line graph and correlogram are checked to understand the trend. The findings of the line graph of the variables Cl, H, R, S, T1, T2, and W are given below (Fig. 1).

From the above graph, it is seen that there is no trend in the above graphs. So, it can be said that those variables are stationary, that is, the mean, variance, and autocovariance are constant through time. To confirm these findings, we also perform a correlogram test and a unit root test.

Correlogram test: The figures of the Correlogram for seven variables are given below. We know that the spike is declining quickly to zero, but the series implies stationary. So, it can be said that the variables are stationary at level (Fig. 2).

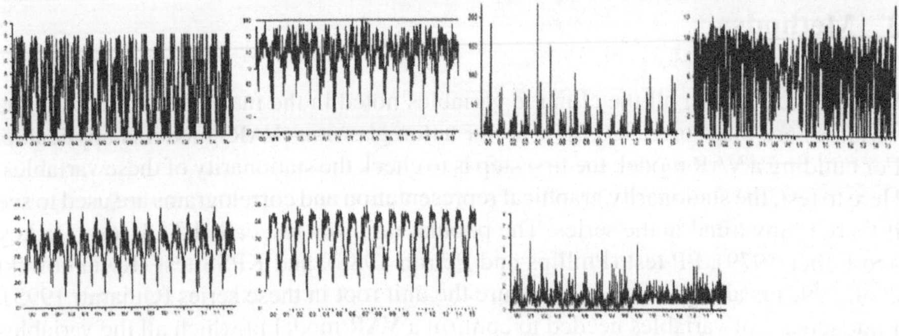

Fig. 1 Line graphs for the variables Cl, H, R, S, T1, T2, and W, respectively (*Source* Data from BARC and the figures are created by the authors using E-Views)

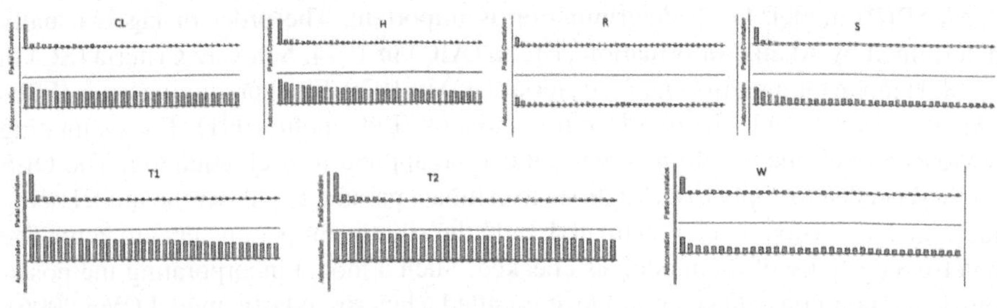

Fig. 2 ACF and PACF graphs for the variables CL, H, R, S, T1, T2, and W, respectively (*Source* Data from BARC and the figures are created by the authors using E-Views)

Unit root test: ADF, PP, and KPSS tests are performed, and the findings are put in Table 1. Here the asterisks mark ***, **, and * indicate that they are significant at 10%, 5%, and 1% level of significance, respectively.

From Table 1, it can be said that there is no unit root in the variables, that is, the variables are stationary.

Table 1 Table for ADF, PP, and KPSS tests for the time series data taken from BARC, and the results are created by the authors

Variables	ADF	PP	KPSS
Cl	-3.431226^{*}	-2.861811^{**}	0.347000^{***}
H	-2.861812^{**}	-2.861811^{**}	0.347000^{***}
R	-2.566956^{***}	-3.431224^{*}	0.463000^{**}
S	-3.431225^{*}	-2.861811^{**}	0.739000^{*}
T1	-3.430516^{*}	-2.861498^{**}	0.463000^{**}
T2	-2.566788^{***}	-3.430516^{*}	0.347000^{***}
W	-3.431226^{*}	-2.566956^{**}	0.463000^{**}

Checking endogeneity: Pairwise Granger Causality tests were performed to check the endogeneity of the variables (Granger, 1969). For the five lag length, the causality is tested. At 1st lag, we found that H, Cl, and T1 do not Granger Cause Cl, T2, and R and also R does not Granger Cause Cl, H, and W. Also S does not Granger Cause H and W, respectively. At lag 2 and 3, the same results are found for both lags' technique where H and S do not Granger Cause Cl and W. Also, R does not Granger Cause Cl, S, and W, respectively. At lag length 4, R does not Granger Cause Cl, S, and W, respectively, and S does not Granger Cause W. At last for lag length 5, we found that R and S do not Granger Cause S and W, respectively. Besides these relations, Granger Cause exists in all other pairs among climatic variables.

From our finding, the unidirectional causality is given as.

$$\text{Cl} \Longrightarrow \text{H},\ \text{Cl} \Longrightarrow \text{R},\ \text{T2} \Longrightarrow \text{Cl},\ \text{H} \Longrightarrow \text{R},\ \text{R} \Longrightarrow \text{T1},\ \text{W} \Longrightarrow \text{R},\ \text{W} \Longrightarrow \text{S},\ \text{S} \Longrightarrow \text{R}.$$

The bidirectional causality is also found as.

$$\text{S} \Longleftrightarrow \text{Cl},\ \text{T1} \Longleftrightarrow \text{Cl},\ \text{W} \Longleftrightarrow \text{Cl},\ \text{T1} \Longleftrightarrow \text{H},\ \text{T2} \Longleftrightarrow \text{H},\ \text{W} \Longleftrightarrow \text{H},\ \text{T2} \Longleftrightarrow \text{R},$$
$$\text{T1} \Longleftrightarrow \text{S},\ \text{T2} \Longleftrightarrow \text{T1},\ \text{W} \Longleftrightarrow \text{T1},\ \text{W} \Longleftrightarrow \text{T2}.$$

Selection of order (P): From Table 2, it is seen that at $P = 3$ the change in the AIC value compared to $P = 4$ is less than 0.04 and also the SC value at $P = 3$ is the smallest among all the values of P after lag 3. Here the lowest value of SC at lag $p = 3$ is chosen. Then VAR (3) model is fitted for estimating $[7 + 49 * 3] = 154$ parameters including intercept terms.

Estimation of parameters: Here the parameters are estimated for VAR (3) models. The coefficients are calculated by E-views and are displayed in Table 3.

The equations ordered by variables are given as follows:

$$\begin{aligned}
\text{CL} = {} & \text{C}(1) * \text{CL}(-1) + \text{C}(2) * \text{CL}(-2) + \text{C}(3) * \text{CL}(-3) + \text{C}(4) * \text{H}(-1) \\
& + \text{C}(5) * \text{H}(-2) + \text{C}(6) * \text{H}(-3) + \text{C}(7) * \text{R}(-1) + \text{C}(8) * \text{R}(-2) \\
& + \text{C}(9) * \text{R}(-3) + \text{C}(10) * \text{S}(-1) + \text{C}(11) * \text{S}(-2) + \text{C}(12) * \text{S}(-3) \\
& + \text{C}(13) * \text{T1}(-1) + \text{C}(14) * \text{T1}(-2) + \text{C}(15) * \text{T1}(-3) + \text{C}(16) * \text{T2}(-1) \\
& + \text{C}(17) * \text{T2}(-2) + \text{C}(18) * \text{T2}(-3) + \text{C}(19) * \text{W}(-1) + \text{C}(20) * \text{W}(-2) \\
& + \text{C}(21) * \text{W}(-3) + \text{C}(22)
\end{aligned}$$

$$\begin{aligned}
\text{H} = {} & \text{C}(23) * \text{CL}(-1) + \text{C}(24) * \text{CL}(-2) + \text{C}(25) * \text{CL}(-3) + \text{C}(26) * \text{H}(-1) \\
& + \text{C}(27) * \text{H}(-2) + \text{C}(28) * \text{H}(-3) + \text{C}(29) * \text{R}(-1) + \text{C}(30) * \text{R}(-2) \\
& + \text{C}(31) * \text{R}(-3) + \text{C}(32) * \text{S}(-1) + \text{C}(33) * \text{S}(-2) + \text{C}(34) * \text{S}(-3) \\
& + \text{C}(35) * \text{T1}(-1) + \text{C}(36) * \text{T1}(-2) + \text{C}(37) * \text{T1}(-3) + \text{C}(38) * \text{T2}(-1) \\
& + \text{C}(39) * \text{T2}(-2) + \text{C}(40) * \text{T2}(-3) + \text{C}(41) * \text{W}(-1) + \text{C}(42) * \text{W}(-2) \\
& + \text{C}(43) * \text{W}(-3) + \text{C}(44)
\end{aligned}$$

Table 2 AIC and SC values for the VAR model at the level for the time series data taken from BARC

Lag	AIC	SC	NPTEST	NOAAEP
1	29.47349	29.53438	56	6209
2	29.31249	29.42666	105	6208
3	**29.25239***	**29.41984***	154	6207
4	29.21465	29.43538	203	6206
5	29.18512	29.45913	252	6205
6	29.1692	29.49649	301	6204
7	29.15797	29.53854	350	6203
8	29.15663	29.59048	399	6202
9	29.15261	29.63974	448	6201
10	29.15568	29.69609	497	6200
11	29.15498	29.74867	546	6199
12	29.15391	29.80088	595	6198
13	29.15475	29.855	644	6197
14	29.15672	29.91026	693	6196
15	29.16137	29.96819	742	6195
16	29.16715	30.02724	791	6194
17	29.17128	30.08466	840	6193
18	29.17834	30.14499	889	6192
19	29.1794	30.19933	938	6191
20	29.18363	30.25685	987	6190

Source Created by the authors using E-Views and R

$$
\begin{aligned}
R = {} & C(45)*CL(-1) + C(46)*CL(-2) + C(47)*CL(-3) + C(48)*H(-1) \\
& + C(49)*H(-2) + C(50)*H(-3) + C(51)*R(-1) + C(52)*R(-2) \\
& + C(53)*R(-3) + C(54)*S(-1) + C(55)*S(-2) + C(56)*S(-3) \\
& + C(57)*T1(-1) + C(58)*T1(-2) + C(59)*T1(-3) + C(60)*T2(-1) \\
& + C(61)*T2(-2) + C(62)*T2(-3) + C(63)*W(-1) + C(64)*W(-2) \\
& + C(65)*W(-3) + C(66)
\end{aligned}
$$

$$
\begin{aligned}
S = {} & C(67)*CL(-1) + C(68)*CL(-2) + C(69)*CL(-3) + C(70)*H(-1) \\
& + C(71)*H(-2) + C(72)*H(-3) + C(73)*R(-1) + C(74)*R(-2) \\
& + C(75)*R(-3) + C(76)*S(-1) + C(77)*S(-2) + C(78)*S(-3) \\
& + C(79)*T1(-1) + C(80)*T1(-2) + C(81)*T1(-3) + C(82)*T2(-1) \\
& + C(83)*T2(-2) + C(84)*T2(-3) + C(85)*W(-1) + C(86)*W(-2) \\
& + C(87)*W(-3) + C(88)
\end{aligned}
$$

Table 3 Estimated parameters of VAR (3) models for seven climatic variables

	CL	H	R	S	T1	T2	W
CL(−1)	0.63057*	0.616786*	1.000562*	−0.37196*	−0.29159*	0.092473*	0.024943*
CL(−2)	−0.00469	−0.2760*	−0.23533	0.145854*	0.143154*	0.067986*	−0.00599
CL(−3)	0.07529*	−0.24949*	0.429685*	0.062872*	0.091703*	0.000294	0.007817
H(−1)	0.02370*	0.653434*	0.508215*	−0.03312*	−0.00253	0.07281*	−0.00922*
H(−2)	0.002915	0.065508*	−0.30388*	−0.00907	−0.01135	0.010531*	0.00412*
H(−3)	0.003212	0.120181*	−0.03152	−0.02291*	−0.03321*	−0.0224*	0.002418
R(−1)	0.001277	−0.00805	0.142976*	0.002222	0.005472*	0.009885*	−0.00032
R(−2)	−0.00276	−0.02103*	0.03046*	0.009289*	0.010921*	0.003772*	−0.00148*
R(−3)	−2.78E−05	0.005687	−0.00075	0.001581	0.004096	0.007006*	−0.00169*
S(−1)	0.003692	0.039949	−0.0537	0.347517*	−0.04341*	−0.00693	−0.00438
S(−2)	0.021269	0.039597	−0.18498*	0.085242*	−0.01664	0.031849*	−0.00052
S(−3)	−0.00267	0.002548	0.111295	0.089507*	0.021254	0.01484	−0.00518
T1(−1)	0.08538*	0.038155	0.331269*	−0.11679*	0.588069*	0.407356	0.003237
T1(−2)	−0.0241	−0.12909	−0.15567	0.008031	0.154138*	0.024383*	0.012996
T1(−3)	0.006858	−0.06327	0.207513	−0.02952	0.017051	−0.07989	0.006184
T2(−1)	0.002702	−0.04	−0.44737*	0.104469*	0.129976*	0.488347*	0.001786
T2(−2)	0.013673	−0.01197	0.043388	0.005096	0.002655	0.082654*	−0.00359
T2(−3)	0.012687	0.201711*	0.092867	0.009512	0.015955	0.122751*	−0.0072
W(−1)	0.23492*	0.843838*	2.203891*	−0.3651*	−0.6101*	−0.28443*	0.468259*
W(−2)	0.02075	−0.26715	−0.40746	0.019034	0.375515*	0.160155*	0.020375
W(−3)	−0.04903	−0.233*	−0.18555	0.004844	0.189571*	0.031837	0.057857*

(continued)

Table 3 (continued)

	CL	H	R	S	T1	T2	W
C	4.47968*	13.4048*	−20.9455*	10.94998*	8.622162*	−10.3015*	0.143586
R-squared	0.675626	0.718057	0.220249	0.364684	0.849369	0.945158	0.323207
Adj. R-squared	0.674525	0.7171	0.217601	0.362527	0.848857	0.944972	0.320909
Akaike AIC	3.684219	6.107511	7.548678	4.665202	3.976512	3.541549	1.359521
Schwarz SC	3.708084	6.131377	7.572544	4.689068	4.000378	3.565414	1.383387

The star marks with red color denote the significant values at a 5% level of significance. *Source* All the results are created by the authors **using E-Views and R**

$$
\begin{aligned}
T1 = {} & C(89)*CL(-1) + C(90)*CL(-2) + C(91)*CL(-3) + C(92)*H(-1) \\
& + C(93)*H(-2) + C(94)*H(-3) + C(95)*R(-1) + C(96)*R(-2) \\
& + C(97)*R(-3) + C(98)*S(-1) + C(99)*S(-2) + C(100)*S(-3) \\
& + C(101)*T1(-1) + C(102)*T1(-2) + C(103)*T1(-3) + C(104)*T2(-1) \\
& + C(105)*T2(-2) + C(106)*T2(-3) + C(107)*W(-1) + C(108)*W(-2) \\
& + C(109)*W(-3) + C(110)
\end{aligned}
$$

$$
\begin{aligned}
T2 = {} & C(111)*CL(-1) + C(112)*CL(-2) + C(113)*CL(-3) + C(114)*H(-1) \\
& + C(115)*H(-2) + C(116)*H(-3) + C(117)*R(-1) + C(118)*R(-2) \\
& + C(119)*R(-3) + C(120)*S(-1) + C(121)*S(-2) + C(122)*S(-3) \\
& + C(123)*T1(-1) + C(124)*T1(-2) + C(125)*T1(-3) + C(126)*T2(-1) \\
& + C(127)*T2(-2) + C(128)*T2(-3) + C(129)*W(-1) + C(130)*W(-2) \\
& + C(131)*W(-3) + C(132)
\end{aligned}
$$

$$
\begin{aligned}
W = {} & C(133)*CL(-1) + C(134)*CL(-2) + C(135)*CL(-3) + C(136)*H(-1) \\
& + C(137)*H(-2) + C(138)*H(-3) + C(139)*R(-1) + C(140)*R(-2) \\
& + C(141)*R(-3) + C(142)*S(-1) + C(143)*S(-2) + C(144)*S(-3) \\
& + C(145)*T1(-1) + C(146)*T1(-2) + C(147)*T1(-3) + C(148)*T2(-1) \\
& + C(149)*T2(-2) + C(150)*T2(-3) + C(151)*W(-1) + C(152)*W(-2) \\
& + C(153)*W(-3) + C(154)
\end{aligned}
$$

From the above seven models, only the significant parameters are kept because they caused these variables. Then the residuals are checked. The standard residuals are also checked as if they lie between 0 ± 3. Heteroscedasticity problems are also checked if there be any. Also, intervention analysis is performed where it is necessary. The fitted models for cloud coverage (CL) is GVAR with GARCH (2,1) and intervention, GVAR with GARCH (3,1) and intervention for Humidity (H), GVAR with GARCH (2,1) and intervention for Maximum Temperature (T1), GVAR with ARCH (2) and intervention for Minimum Temperature (T2), GVAR with intervention for Sunshine (S), ARIMA (1,0,1) with GARCH (1,1) for Rainfall (R), and ARIMA (2,0,2) for wind speed (W).

Here rainfall (R) and Wind speed (W) are a bit different because the observation of these two variables fluctuate much and so GVAR did not work here. That is why separately ARIMA is used for these two variables and a good fit is observed.

5 Discussion

Results of four studies including both univariate and multivariate data are displayed in Table 4 below. ARIMA and SARIMA are used for univariate data, while the VAR

Table 4 Comparison of different results among previous and present studies

Authors	Population/data	Fitted model	Dependent variable	AIC and BIC
Shahin et al. (2014)	Climate data	VAR (6)	Temperature	3.425 and 3.8102
			Rain	11.89 and 12.278
			Wind Speed	0.540 and 0.9260
			Humidity	5.621 and 6.0066
			Cloud	2.338 and 2.7236
			Bright sunshine	2.678 and 3.0632
Ferdous et al. (2014)	Climate data	SARIMA $(2, 0, 1)(1, 0, 2)_{12}$	Temperature	1.29 and 1.35
Ahmed and Hossain (2018)	Climate data	VAR (9) SARIMA	MAXT	1319.9 and 1344.5
			MINT	1592.56 and 1645.80
			HUM	2296.30 and 2329.05
			CLOUD	1044.04 and 1085.00
Present study	Climate data	GVAR with GARCH (2,1) and intervention GVAR with GARCH (3,1) and intervention ARIMA (1,0,1) with GARCH (1,1) GVAR with intervention GVAR with GARCH (2,1) and intervention GVAR with ARCH (2) and intervention ARIMA (2,0,2)	Cloud	3.173 and 3.187
			Humidity	4.084 and 4.095
			Rainfall	1.132 and 1.137
			Bright sunshine	3.594 and 3.597
			Maximum temperature	3.251 and 3.259
			Minimum temperature	3.187 and 3.193
			Wind speed	−3.061 and −3.066

Source Past data by different authors (their references are given) and the present study is created by the authors using E-Views

model is used for multivariate data. The present study is based on daily basis multivariate data where a generalized VAR model is used for analysis. Other multivariate studies are based on monthly data. More and above, intercorrelation among the variables has been considered in the present study resulting in less chance of information loss. Akaike Information criteria (AIC) and Schwarz criteria (SC) displayed in Table

Table 5 Table for bias proportion from the forecasted data (data was taken from the final model that was built in this study)

Variables	In sample forecast	Out-of-sample forecast
Cloud coverage	0.000143	0.000076
Humidity	0.000052	0.000023
Maximum temperature	0.01514	0.000534
Minimum temperature	0.001446	0.000300
Bright sunshine	0.0141	0.0139
Rainfall	0.00005	0.000002
Wind speed	0.013791	0.00000015

Source Created by the authors using E-Views and R

4 confirm this claim. Considering the facts mentioned, it may be claimed that the present study is more efficient and reliable than those of earlier studies for forecasting climate variables.

Forecasting: To forecast, the stability of the models has been checked. The variance proportions of the variables were compared. By definition, the out-of-sample variance proportion must be smaller than the in-sample variance proportion to get a stable model. The variance proportions of the variables are given in Table 5.

From Table 5, it is seen that the out-of-sample forecast errors are smaller than the in-sample forecast error. Some are too close. From this finding, it can be said that the models are more or less stable.

6 Conclusion

Climate change is now a burning issue around the world and even in Bangladesh. The Rajshahi district is situated in the north-western region of Bangladesh. Here the weather is a bit different. In the summer, people feel so much heat and bright sunshine. In the rainy season, the lowest rainfall has been recorded among the districts of the country. In the winter season, the temperature comes down. So, we may consider an interrelation among all these climatic variables. After performing the Granger causality test, it is found that the variables under study are interrelated. As the variables are declared from the granger causality test to be interrelated, a VAR or Vector autoregression model has been built. First of all, the stationarity of the variables was checked and found every variable is stationary at level. The unidirectional and bidirectional causality are checked by the Granger causality test. It is found from the test that some variables are causing each other, and some have caused only the others. So, the interrelationship among them is confirmed. Then a VAR model is confirmed to build. Lag 3 was selected upon the lowest value of AIC and SC. Then VAR (3) model was selected. After getting the seven VAR models for different seven variables, every model has taken separately and GARCH was applied to remove the heteroscedastic problem from the models. Also, intervention analysis is used where necessary. The stability of these models is checked and it is found that out-of-sample forecast errors are smaller than in-sample forecast errors, which defines that the fitted models were stable and they are suitable for further forecasting. We forecasted from 2018 to 2022 with these stable models and it was found that the Maximum Temperature (T1), Humidity (H), Bright Sunshine (S), and Wind Speed (W) might be shown an upward trend and Minimum Temperature (T2), Rainfall (R) and Cloud Coverage (Cl) might be shown decreasing trend from 2018 to 2022.

Natural calamities like floods, cyclones, northwester, land erosion, lowering the water level, etc. are now affecting the world. Human as well as animal's life is at stake. If we can know the situation before it happens, then we can make people aware of those disasters and can take precautionary steps to minimize the loss; policymakers can think about the whole situation, which must lead them to think about alternatives.

Acknowledgements We express our sincere thanks to the Ministry of Education (MoE), Bangladesh for their financial support of this research work. We also express our deep gratitude to the reviewers for their constructive comments and suggestions. Our deep gratitude goes to all the concerned persons of the Springer as well as of the Department of Statistics, the University of Rajshahi for accepting this manuscript for publication.

References

Ahmed, A., & Hossen, M. J. (2018). Spatial and temporal variations of temperature in Bangladesh: An analysis from 1950 to 2012. *Oriental Geographer, 58*(1), 1–22.

Akaike, H. (1969). Fitting autoregressive models for prediction. *Annals of the Institute of Statistical Mathematics, 21*, 243–247.

Akaike, H. (1974). A new look at the statistical model identification. *IEEE Transactions on Automatic Control AC, 19*, 716–723.

Box, G. E. P., & Tiao, G. C. (1975). Intervention analysis with applications to economic and environmental problems. *Journal of the American Statistical Association, 70*(349), 70–79. http://links.jstor.org/sici?sici=0162-1459%28197503%2970%3A349%3C70%3AIAWATE%3E2.0.CO%3B2-4.

Cutter, S. L., & Finch, C. (2008). Temporal and spatial changes in social vulnerability to natural hazards. doi:https://doi.org/10.1073/pnas.0710375105.

Dilley, M., Chen, R. S., Deichmann, U., Lerner-Lam, A. L., & Arnold, M. (2005). *Natural disaster hotspots: A global risk analysis*. Washington, DC: World Bank. © World Bank. https://openknowledge.worldbank.org/handle/10986/7376

Dickey, D. A., & Fuller, W. A. (1979). Distribution of the estimators for autoregressive time series with a unit root. *Journal of the American Statistical Association., 74*, 427–431.

Ferdous, A., Shahin, M. A., & Ali, M. A. (2014). Time series model building and forecasting on maximum temperature data in the book*GIS visualization of climate change and prediction of human responses* (pp.79–91). Springer, New York.

Granger, C. W. J. (1969). Investigating causal relations by ecomometric methods and cross-spectral methods. *Econometrica, 37*, 424–438.

Gujarati, D. N. (1993). *Basic econometrics* (3rd ed.) (pp: 835–854). New York: McGraw-Hill International.

Gujarati, D. N. (1995). *Basic econometrics* (4th ed.). United State Military Academy.

Hallegatte, S., & Przyluski, V. (2010). the economics of natural disasters: Concepts and methods. Policy Research working paper; no. WPS 5507. World Bank. © World Bank. https://openknowledge.worldbank.org/handle/10986/3991

Hannan, E. J., & Quinn, B. G. (1979). The determination of the order of an autoregression. *Journal of the Royal Statistical Society: Series B, 41*, 190–195.

Khan, M. A. R., & Ali, M. A. (2003). VAR modeling with mixed series. *International Journal of Statistical Sciences., 2*, 19–25.

Kleiber, W., Katz, R. W., & Rajagopalan, B. (2013). Daily minimum and maximum temperature simulation over complex terrain. *Annals of Applied Statistics, 7*(1), 588–612. https://doi.org/10.1214/12-AOAS602

Kwiatkowski, D., Phillips, P. C. B., Schmidt, P., & Shin, Y. (1992). Testing the null hypothesis of stationary against the alternative of a unit root. *Journal of Econometrics, 54*, 159–178.

Lütkepohl, H. (1991). *Introduction to multiple time series analysis.* Berlin: Springer.

Phillips, P. C. B., & Perron, P. (1988). Testing for a unit root in time series regression. *Biometrika, 75*, 335–346.

Schwarz, G. (1978). Estimating the dimension of a model. *Annals of Statistics, 6*, 461–464.

Shamsina, & Hossain, B. (2011). Modeling climatic variables using time series analysis in arid and semi arid region. African Journal of Agricultural Research, june, 2014, *African Journal of Agricultural Research, 9*(26), 2018-2027. doi:https://doi.org/10.5897/AJAR11.1128.

Shahin, A., Ali, M. A., & Ali, A. B. M. S. (2014). Vector autoregression (VAR) modeling and forecasting of temperature, humidity, and cloud coverage, in the book *GIS visualization of climate change and prediction of human responses* (pp.29–51). New York: Springer.

Wei, W. W. S. (2006). *Time series analysis: Univariate and multivariate methods* (2nd ed.). Pearson Addison Wesley.

Wei, W. W. S. (1990). Time series analysis: Univariate and multivariate methods. *International Journal of Forecasting, 7*(3), November 1991, 389–390. http://www.sciencedirect.com/science/article/pii/0169-2070(91)90015-N.

Experimental Designs for fMRI Studies in Small Samples

Bikas K. Sinha

Abstract Functional Magnetic Resonance Imaging (fMRI) is a technology for studying how our brains respond to mental stimuli. At the design stage, one is interested in developing the best sequence of mental stimuli for collecting the most informative data in order to render the most precise inference about the 'unknown parameters' under an assumed statistical model. The simplest such model incorporates linear relation between mean response and the parameters describing the effects of the stimuli, applied at regularly spaced time points during the study period. In this paper, we introduce the linear model and discuss estimation issues and related concepts such as 'orthogonality' and 'balance'.

Keywords fMRI · Linear model · Spring balance · Bias · h-parameters · Estimability · Balanced structure · Orthogonality · Information matrix · Generalized variance

1 Introduction

The key reference to this article is Cheng and Kao (2015) who carried out an investigation on the scope of study of optimal experimental designs in the context of fMRI studies.

The brain functions of the experimental subjects are captured through response profiles at a number of instances. Each subject experiences the onset of a stimulus at an instant if the stimulus is active [recognized by code 1] at that instant; otherwise, the subject is at a resting state [recognized by code 0] at that instant. Each instant is defined as a compact duration of four seconds. At any instant, the brain voxel captures the cumulative effects of a constant parameter θ and the h-parameters at the current instant as well as at each of the immediate past ordered (K − 1) instants—for

This chapter draws some material from the author's previous work presented in a conference and published in the Proc. Volume of the said Conference. The author has received necessary permission from the pub-lisher of the said Conference Proceedings Volume.

B. K. Sinha (✉)
Indian Statistical Institute, Kolkata, India

B. K. Sinha et al. (eds.), *Data Science and SDGs*,
https://doi.org/10.1007/978-981-16-1919-9_8

some K—whenever there has been an onset of active stimulus [coded by 1] at any of these instances. The reader familiar with the concept of 'carry-over effects' in the context of Repeated Measurement Designs [RMDs] or Cross-Over Designs will find a similarity in the model description (Vide Shah and Sinha 1989).

A more general scenario exhibits itself in terms of different stages of activation of the brain stimuli, rather than just being 'active'—as coded by '1' in the above. We refer to Kao et al. (2008) for this and related considerations.

The mean model formulation is developed as follows. An experimental design of length n, say D_n, is a description of a sequence of 0 s and 1 s of total length n. For example, for $n = 7$, the following describes a seven-point design: $D_7 = [0, 1, 1, 0, 1, 0, 0]$. The utility of the suggested design D_7 is described below. For any n, D_n is very much like D_7. The linear model to be described below is developed as a 'circular' model—a well-known consideration in the context of RMDs or Cross-Over Designs (Vide Kunert 1984 or Shah and Sinha 1989). To visualize a circular model, the same sequence (describing D_7) is used as a 'dummy' sequence and this is described as follows:

$< 0, 1, 1, 0, 1, 0, 0 >$	$\rightarrow$	$[0, 1, 1, 0, 1, 0, 0]$
Dummy Sequence	**followed by**	**Data-generating Sequence**

There are seven data/time points and as such we observe y_1 to y_7 corresponding to the seven time points in the data-generating sequence [0, 1, 1, 0, 1, 0, 0]—going from left to right. In the terminology of RMDs or Cross-Over Designs, for the fi time point, the 'direct effect' [denoted by h_1] is to be captured along with the 'carry-over effects' [h_2,h_3, ….] of the preceding time points as described in the dummy Sequence—from right to left. Although, at each data point, only if the stimulus is active [denoted by 1], the corresponding h-parameter will be considered. Lastly, for n data/time points, we can incorporate at the most n 'parameters' including the constant parameter θ. This implies that we can incorporate in the model at the most $(n - 1)$ h-parameters. Otherwise, identifiability/estimability issues creep in. In terms of K, it means that we assume—to start with—that $K \leq (n - 1)$.

We start with the following Table 1 describing the linear model underlying the design D_7. We assume $K = 6$.

Remark 1: We find remarkable similarity of this model to what is generated in the context of 'biased spring balance weighing designs' (Vide Raghavarao 1971 or Shah and Sinha 1989). Naturally, the θ-parameter represents the bias component. The co-efficient matrix $\mathbf{X} = ((x_{ij}))$ consists of 0 s and 1 s. It may be noted that in the above, the **X**- matrix is shown in the reverse order. Multiplication by a permutation matrix **P** will bring it to the right/standard order. Finally, the linear model ($\mathbf{Y}$, $\mathbf{X}^{(*)}\beta$, $\sigma^2\mathbf{I}$) is obtained as usual, where $\mathbf{X}^{(*)} = (\mathbf{1}, \mathbf{PX})$ and $\beta = (\theta, h1, h2, \ldots)'$.

Remark 2: We recall the implication of the 'circular model' in this context. The data-generating sequence is the same as the dummy-sequence on which the circular model is built. Another implication is that the columns $\mathbf{h}_1$, $\mathbf{h}_2$, ... are circular in nature. That is, the columns of matrix **X** are circular in nature. For a non-circular

Table 1. Linear model with positional carry-over effects

S1. no	h_6	h_5	h_4	h_3	h_2	h_1	y	Mean model
1	1	0	1	0	0	0	y_1	$\theta + h_4 + h_6$
2	0	1	0	0	0	1	y_2	$\theta + h_1 + h5$
3	1	0	0	0	1	1	y_3	$\theta + h_1 + h_2 + h_6$
4	0	0	0	1	1	0	y_4	$\theta + h_2 + h_3$
5	0	0	1	1	0	1	y_5	$\theta + h_1 + h_3 + h_4$
6	0	1	1	0	1	0	y_6	$\theta + h_2 + h_4 + h_5$
7	1	1	0	1	0	0	y_7	$\theta + h_3 + h_5 + h_6$

design/model, the carry-over effects are to be determined separately—depending on the nature of 1 s and 0 s—for each incoming unit/patient.

In case it is believed that there are only $K^*[<K]$ h-parameters in the model, the understanding is that the initial set of K^*h-parameters viz., h_1,h_2,..., h_{K*} is important and the rest can be ignored from the mean model. For $K^* = 4$, the model expectations of successive responses corresponding to the above design would be.

$$\theta + h_4, \theta + h_1, \theta + h_1 + h_2, \theta + h_2 + h_3, \theta + h_1 + h_3, \theta + h_2 + h_4, \theta + h_3.$$

Naturally the above design with $n = 7$ instances [for experimentation] now turns out to be an over-saturated design when only $K^* = 4$ h-parameters are believed to be present. In this case, we might curtail the experiment from D_7 to D_5 since there are five parameters, including the common parameter θ. Use of D_5: [0, 1, 1, 0, 1] provides for the mean model the expressions.

$$\theta + h_2 + h_4, \theta + h_1 + h_3, \theta + h_1 + h_2 + h_4, \theta + h_2 + h_3, \theta + h_1 + h_3 + h_4.$$

On the other hand, use of $D_{alt.5}$: [1, 0, 1, 0, 0] provides for the mean model the expressions.

$$\theta + h_1 + h_4, \theta + h_2, \theta + h_1 + h_3, \theta + h_2 + h_4, \theta + h_3.$$

Note that in both cases, we have taken due consideration of the circular nature of the sequence in working out the mean models. A natural question would be to search out the difference if any, between the two D_5 designs. Popular optimality criteria rest on the computation of the 'information matrix' for the h-parameters based on the Gauss-Markov Model, assuming homoscedastic errors with mean 0 and variance σ^2. Minimization of the generalized variance [computed as reciprocal of the determinant of the information matrix] is an acceptable criterion for the choice of the best design. This is the so-called D-optimality criterion (Vide Shah and Sinha 1989).

2 Estimability Issues

For a given n, we can incorporate a maximal set of $(n - 1)$ h-parameters. In other words, we can develop the full model with θ and additional $(n - 1)$ h-parameters. Naturally, the response vector $\mathbf{Y}$ of dimension $n \times 1$ will come under the standard Gauss-Markov linear model described earlier.

Suppose the design sequence Dn has r 1 s and $(n - r)$ 0 s. Since the choice of the design is circular in nature, it easily follows that each column sum of $\mathbf{X}$ is r.

We have already introduced the 'design matrix' $\mathbf{X}^{(*)} = (\mathbf{1}, \mathbf{PX})$ and the underlying parameters $\beta = (\theta, h_1, h_2, \ldots)'$. For a given design D_n, when there are $K[\leq (n-1)]$ h-parameters viz., h_1,h_2,..., h_k in the model, the h-parameters are all estimable iff Rank $(\mathbf{X}^{(*)}) = 1 + K$, where $\mathbf{X}^{(*)}$ is based on K column vectors corresponding to the K h-parameters, in addition to the column vector $\mathbf{1}$. The 'if' part is easy to see. On the other hand, if all the h-parameters are estimable, θ is trivially so based on any single observation and hence the rank condition is satisfied.

In the above example, for $K = 6$, the design D_7 provides estimability of all the h-parameters. Again, for $n = 6$, the design sequence $D_6 = [1, 0, 1, 0, 1, 0]$ provides estimability of the h-parameters only for $K = 2$ and fails for $K = 3, 4, 5$.

3 Choice of N and D_n for Given K^*

It is a priori known that $h_1, h_2, \ldots, h_{K^*}$ are the only h-parameters present in the mean model. Therefore, we need $n \geq (1 + K^*)$ design points, and the choice of D_n must be such that the formation of $\mathbf{X}$ enables one to ensure rank condition. Would any circulant of length n with an arbitrary row/column total ensure the rank condition?

Consider the design D_n: [1, 0, 0..., 0, 0].

The model expectations of the successive responses ys are based on the following realizations and are given below.

[1, 0,..., 0, 0] *dummy sequence followed by* [1, 0,..., 0, 0] *data-gathering sequence.*

Model expectations are given by $[\theta + h_1, \theta + h_2, \theta + h_3, \ldots, \theta + h_{(n-1)}, \theta]$. It is evident that this class of designs forms an admissible class whatever be $n \geq (1 + K^*)$, K^* being the number of h-parameters in the model.

Further to this, for an arbitrary $n = 1 + K^* + E$ [E standing for the excess number of data points], it follows that based on this type of formation of D_n,

(*i*) $\hat{\theta} = \bar{y}^*$; (*ii*) $\hat{h}_i = y_i - \hat{y}^*$; $i = 1, 2, \ldots$, (*iii*) $Var\left(\hat{h}_i\right) = (1 + c)\sigma^2$; $i = 1,2, \ldots$, where $\bar{y}^* = \sum_{i>K^*} y_i/(n - K^*)y_i/(n - K^*)$, $c = 1/(n - K^*)$.

Moreover, we also have $\bar{V}\left(\hat{h}\right) = (1 - c)\sigma^2$.

Remark 3: It must be noted that for any n, a design D_n cannot be formed using exclusively 1 s or 0 s. In other words, the patient cannot be in a resting or in an active state all through the time duration for the collection of data. A mixture of the two states is called for.

Remark 4: For the given K^* [the number of non-null h-parameters], if D_n is an admissible n-point design, then one can always add a few 0's and 1's in 'suitable' positions so that the 'extended' design is also admissible. The point to be noted is that we are dealing with a circular model. With the addition of a few data points, this circular property will change the model expectations and therefore, admissibility/estimability may not be taken for granted. Here is an example of this effect.

Consider $K^* = 3$ and $D_5 = [0, 1, 1, 0, 1]$ as taken above. The h-parameters are captured by the column vectors in the following Table 2.

It is easy to verify that all the three h-parameters and θ are estimable. If we extend this design to $D_6^* = [0, 1, 1, 0, 1; 0]$, by the circular nature of the model, the h-parameters are now captured by the column vectors in the following Table 3.

Estimability is still ensured for the design D_6^*. If, instead, we extend D_5 to $D_6^{**} = [0, 1, 1, 0, 1; 1]$, then the column vectors are given in Table 4.

Table 2 Linear model with positional carry-over effects

Sl. No	h_3	h_2	h_1	y	Mean model
1	0	1	0	y_1	$\theta + h_2$
2	1	0	1	y_2	$\theta + h_1 + h_3$
3	0	1	1	y_3	$\theta + h_1 + h_2$
4	1	1	0	y_4	$\theta + h_2 + h_3$
5	1	0	1	y_5	$\theta + h_1 + h_3$

Table 3 Linear model with positional carry-over effects

Sl. No	$h3$	$h2$	$h1$	y	Mean model
1	1	0	0	y_1	$\theta + h_3$
2	0	0	1	y_2	$\theta + h_1$
3	0	1	1	y_3	$\theta + h_1 + h_2$
4	1	1	0	y_4	$\theta + h_2 + h_3$
5	1	0	1	y_5	$\theta + h_1 + h_3$
6	0	1	0	y_6	$\theta + h_2$

Table 4 Linear model with positional carry-over effects

Sl. No	h_3	h_2	h_1	y	Mean model
1	1	1	0	y_1	$\theta + h_2 + h_3$
2	1	0	1	y_2	$\theta + h_1 + h_3$
3	0	1	1	y_3	$\theta + h_1 + h_2$
4	1	1	0	y_4	$\theta + h_2 + h_3$
5	1	0	1	y_5	$\theta + h_1 + h_3$
6	0	1	1	y_6	$\theta + h_1 + h_2$

It follows that this time, not all the h-parameters are estimable since θ is not estimable. The circular nature of the design/model explains these results. Therefore, the extension cannot be taken for granted.

4 Comparison of Design Sequences

When we address this problem for design sequences of the same length n, there are effectively $2^n - 2$ such comparable sequences barring the two extremes [all 0 s and all 1 s]. The actual number of admissible sequences may be much smaller—depending on the number K of non-negligible h-parameters. Anyway, such a comparison of two admissible sequences may rest on, say the criterion of smaller average variance or smaller generalized variance of the estimated h-parameters. Below we take up the case of the saturated model with $n = 5$, $K = 4$ and compare all available admissible design sequences. It follows that all the 30 design sequences are admissible and these are classified into three distinct types as follows:

Type I: (a) [0, 0, 0, 0, 1]; (b) [0, 0, 0, 1, 1];
(c) [0, 0, 1, 1, 1]; (d) [0, 1, 1, 1, 1]
and all their cyclic permutations—covering 20 design sequences.

Type II: [1, 0, 1, 0, 0]; and its cyclic permutations—covering five design sequences.

Type III: [1, 1, 0, 1, 0] and its cyclic permutations—covering five design sequences.

The computations are routinely carried out and the average variance of estimated h-parameters for all the above choices are given in Table 5.

This exercise suggests that there is a need to examine the status of different available design sequences of the same length. Here we compared the design sequences with respect to the criterion of the smallest average variance. More interesting would be the scenario when a design sequence D_n is 'extended' to D_{n^*} for some $n^* > n$. Under a circular model, the question of admissibility of the extended sequence has to be settled first. And when this obtains, we have to work out average variance and/or generalized variance and compare across different admissible choices of the design

Table 5. Design sequences, variances, and average variances

S1. No	$V(\hat{h}_1)$	$V(\hat{h}_2)$	$V(\hat{h}_3)$	$V(\hat{h}_4)$	Average v ariance
$I(a)$	$2\sigma^2$	$2\sigma^2$	$2\sigma^2$	$2\sigma^2$	$2\sigma^2$
$I(b)$	$4\sigma^2$	$2\sigma^2$	$2\sigma^2$	$4\sigma^2$	$3\sigma^2$
$I(c)$	$4\sigma^2$	$2\sigma^2$	$2\sigma^2$	$4\sigma^2$	$3\sigma^2$
$I(d)$	$2\sigma^2$	$2\sigma^2$	$2\sigma^2$	$2\sigma^2$	$2\sigma^2$
II	$2\sigma^2$	$4\sigma^2$	$4\sigma^2$	$2\sigma^2$	$3\sigma^2$
III	$2\sigma^2$	$4\sigma^2$	$4\sigma^2$	$2\sigma^2$	$3\sigma^2$

sequences—even though the two designs are based on the unequal number of design points. Is it always the case that the latter design fares better than the former when the latter is also admissible? These are non-trivial questions to be settled. We will not pursue this topic any further.

5 Structural Balance in and Orthogonality of a Design Sequence

For $K = 3$ h-parameters, the mean model generally involves expressions such as $\theta + h_i$, $i = 1, 2, 3$; $\theta + h_i + h_j$, and $i \neq j$. The underlying design sequence is said to be 'Structurally Balanced [SBDS]' when frequency counts of $\theta + h_1$, $\theta + h_2$, *and* $\theta + h_3$ are the same and also those of $\theta + h_1 + h_2$, $\theta + h_1 + h_3$, and $\theta + h_2 + h_3$ are the same. In other words, for an SBDS with $K = 3$, in terms of the h-parameters, singletons are equally frequent, and also doubletons are equally frequent. The notion generalizes naturally for higher values of K. When this happens, the information matrix of the h-parameters turns out to be 'completely symmetric', i.e., one with all diagonal elements equal and all off-diagonal elements equal. Specifically, for $K = 3$ and for an SBDS of size n, let f_0, f_1, f_2, *and* f_3 represent, respectively, the frequency counts of θ, each of the singletons, each of the doubletons and finally, that of $\theta + h_1 + h_2 + h_3$. Then $n = f_0 + 3f_1 + 3f_2 + f_3$ and the 4×4 information matrix of the β-parameter vector is given by.

$$[n, f_1 + f_2 + f_3, f_1 + f_2 + f_3, f_1 + f_2 + f_3]; [---, f_1 + 2f_2 + f_3, f_2 + f_3, f_2 + f_3]; [---, ---, f_1 + 2f_2 + f_3, f_2 + f_3]; [---, ---, ---, f_1 + 2f_2 + f_3].$$

It transpires that the information matrix for the h-parameters is completely symmetric. Further, orthogonality of the estimates happens whenever.

$$(f_0 + 2f_1 + f_2)(f_2 + f_3) = (f_1 + 2f_2 + f_3)(f_1 + f_2),$$

meaning thereby that pairwise covariances of estimates of the h-parameters are zeros. Here is an example for $n = 8$, $K = 3$: $D_8 = [1, 1, 1, 0, 1, 0, 0, 0]$. It is structurally balanced with $f_0 = f_1 = f_2 = f_3 = 1$. Hence, the orthogonality condition is trivially satisfied.

It is interesting to note that a design sequence can be structurally unbalanced but still it may turn out to be orthogonal. Here is an example,

$$n = 12, K^* = 4 \ : D_{12} = [1, 01, 1, 1, 1, 0, 0, 0, 1, 0, 0]$$

Whereas each one of the coefficient vectors (involving the h-parameters).

$$[1000]'; [0100]', [0001]'$$

appears exactly once, but $[0010]'$appears twice in the model expectations of the observations. That explains the structural imbalance of the design sequence. In spite of that, it turns out that $V\left(\hat{h}_i\right) = 3; Cov\left(\hat{h}_i, \hat{h}_j\right) = 0; i \neq j = 1, 2, 3, 4.$

Cheng and Kao (2015) found out a design sequence for $n = 20$ and $K = 5$ which is both structurally balanced and orthogonal,

$$D_{20} = [1, 0, 0, 1, 1, 1, 1, 1, 0, 1, 0, 1, 0, 0, 0, 0, 1, 1, 0, 0]$$

A combinatorially challenging problem would be, for a given K, to find out balanced orthogonal design with smallest number n of design sequence. Another related problem would be to ascertain if an orthogonal/structurally balanced design sequence for a given value of K continues to be so for smaller/higher values of K.

6 Use of Hadamard Matrices

It is well-known that there is a close connection between Hadamard Matrices and Weighing Design Matrices [of both types: Chemical Balance and Spring Balance] Vide Raghavarao (1971) and Shah and Sinha (1989). Researchers have examined the possibility of starting with Hadamard Matrices and converting them into ((0, 1)) matrices so as to examine their properties as Design Sequences for fMRI experiments. Once more we re-iterate that we are adopting a circular model through the fMRI experiment. We will not dwell on this topic at all. We refer to serious studies along this direction.

Acknowledgements The author is thankful to Professors Nripes K Mandal and Manisha Pal, Department of Statistics, Calcutta University, for taking keen interest in this topic of research and for fruitful discussions. The author also thanks an anonymous referee for making some useful suggestions in an earlier draft.

References

Cheng, C. S., & Kao, M. H. (2015). Optimal experimental designs for fMRI via circulant biased weighing designs. *The Annals of Statistics, 43*(6), 2565–2587.

Kao, M. H., Mandal, A., Stufken, J. (2008). Optimal design for event-related functional magnetic resonance imaging considering both individual stimulus effects and pairwise contrasts. *Statistics and Applications*, 6(1–2), 235–256.

Kunert, J. (1984). Optimality of balanced uniform repeated measurements de-signs. *The Annals of Statistics, 12*, 10061017.

Raghavarao, D. (1971). *Construction and combinatorial problems in design of experiments*. John Wiley and Sons.

Shah, K. R., & Sinha, B. K. (1989). *Theory of optimal designs*. Springer Lecture Notes in Statistics Series No. 54.

Bioinformatic Analysis of Differentially Expressed Genes (DEGs) Detected from RNA-Sequence Profiles of Mouse Striatum

Bandhan Sarker, Md. Matiur Rahaman, Suman Khan, Priyanka Bosu, and Md. Nurul Haque Mollah

Abstract Bioinformatic analysis is a powerful statistical analysis to investigate the significant genes and their biological information from RNA-sequence (RNA-Seq)-based gene expression profiles. The most differentially expressed genes (DEGs) of mouse striatum with their valuable information may be significantly contributed to the neuroscience research. Two inbred mouse strains, for instance, C57BL/6J (B6) and DBA/2J (D2), in neuroscience research are commonly used, and B6 strain sequences are mostly available. Our study's focus on the identification of significant DEGs of B6 and D2 samples, protein–protein interaction network, to identify their biological functions, molecular pathway analysis, miRNAs-target gene interactions, downstream analysis, and to find out driven genes. Two samples, 10 B6 and 11 D2, were deeply analyzed, which were retrieved from the Gene Expression Omnibus (GEO) database with accession number GSE26024. DESeq2, edgeR, and limma tools were utilized to screen the DEGs somewhere in the range of B6 and D2 samples. DESeq2, edgeR, and limma had identified a total of 736, 757, and 530 DEGs with 37, 48, and 31 up-regulated genes, respectively. Protein–protein interaction network analyses of those DEGs were visualized using a search tool for the Retrieval of Interacting Genes and Cytoscape software. We selected the top 50 high-degree hub DEGs for each of the three methods, and explored 21 common hub genes along with three up-regulated genes Bdkrb2, Aplnr, and Ccl28. To explore the biological insights of these 21 common hub DEGs, Gene Ontology (GO) and KEGG pathway analysis were executed. In downstream analysis, hierarchical and k-means clustering techniques were used, and both the methods clustered Bdkrb2, Aplnr, and Ccl28 genes into the same group. Furthermore, DEGs, specifically the genes Bdkrb2, Aplnr, and Ccl28, are probably the core genes in inbred mouse strains. In conclusion, these genes probably are the biomarkers for further neuroscience research.

B. Sarker · Md. M. Rahaman (✉) · S. Khan · P. Bosu
Department of Statistics, Faculty of Science, Bangabandhu Sheikh Mujibur Rahman Science and Technology University, Gopalganj 8100, Bangladesh

Md. N. H. Mollah
Bioinformatics Laboratory, Department of Statistics, University of Rajshahi, Rajshahi 6205, Bangladesh
e-mail: mollah.stat.bio@ru.ac.bd

B. K. Sinha et al. (eds.), *Data Science and SDGs*,
https://doi.org/10.1007/978-981-16-1919-9_9

Keywords Mouse RNA-Seq profiles · Differential gene expressions · Functional analysis · Molecular pathway analysis · Bioinformatic approaches

1 Introduction

RNA-Seq may be a way to investigate the number and sequences of RNA in a sample. Over the past decade, the revolution of next-generation sequencing has exceedingly produced a greater yield of sequence data at an inferior cost (Van Dijk et al. 2014). Simultaneously, analysis techniques used for inspecting sequence data have emerged (Alioto et al. 2013; Anders et al. 2013; Huber et al. 2015). Among the widespread methods, RNA-Seq is the largest project for analyzing sequence data. Over the past decade, the genome-wide mRNA expression data derivation from cell population has been demonstrated to be useful in many more studies (Soneson and Delorenzi 2013; Bacher and Kendziorski 2016).

Although traditional expression methods existed for analyzing thousands of cells, they sometimes cover or even misinterpret ones of interest. Nowadays, advanced technologies allow us to induce transcriptome-wide large-scale information from cells. This improvement is not simply another progression to enhance expression profiling, yet rather a major development that will empower crucial experiences into biology (Bacher and Kendziorski 2016). The analysis of RNA-Seq data assumes a crucial part to understand the inherent and extraneous cell measures in biological and biomedical exploration (Wang et al. 2019a). To understand biological processes, a more precise understanding of the transcriptome in cells is needed for explicating their role in cellular functions and understanding how differentially expressed genes (DEGs) can promote advantageous or harmful design (Hwang et al. 2018). Appropriate analysis and utilization of the massive amounts of data generated from RNA-Seq experiments are challenging (Pop and Salzberg 2008; Shendure and Ji 2008). However, DEGs detection is one of the most significant efforts in RNA-Seq data analysis. Several methods have been used for identifying DEGs from count RNA-Seq data in bioinformatic analysis based on Poisson and negative binomial distribution. Poisson distribution faces an over-dispersion problem; therefore, the negative binomial distribution is more reliable. In this study, we used three familiar methods (DESeq2, edgeR, and limma) to follow negative binomial distribution for examining DEGs, and we are going to discuss the fundamental principles of bioinformatic techniques, focusing on concepts important in the analysis of RNA-Seq mouse striatum data.

Multiple brain regions based on different inbred mouse strains gene expression profiles have been established previously (Sandberg et al. 2000; Hovatta et al. 2005). The distinct opioid-related phenotype has been studied by gene expression profiling in the mouse striatum (Korostynski et al. 2006). Strain reviews exhibited that affectability to morphine is an unprecedented degree reliant on hereditary determinants. In our study, we performed bioinformatic analysis on gene expression profiles of mouse striatum and chose two samples, C57BL/6J and DBA/2J. DESeq2, edgeR, and limma detected the DEGs and took the top 50 DEGs from each. From these genes, we determined common hub DEGs, and performed GO annotation and KEGG pathway analysis. For common hub DEGs, miRNA–mRNA network is constructed. After that, downstream analysis is also carried out to find the driven genes. Therefore, the bioinformatic approach paved the way for the investigation of genes from RNA-Seq profiles of mouse striatum that can be contributed further to molecular research in neuroscience.

2 Materials and Methods

We analyzed RNA-Seq read count data of mouse striatum. The following flow chart shown in Fig. 1 describes the steps of bioinformatic analysis of the data set used in this study.

2.1 RNA-Seq Data Collection

We downloaded gene expression profile GSE26024 from the Gene Expression Omnibus (GEO, https://www.ncbi.nlm.nih.gov/geo/query/acc.cgi?acc=GSE26024) (Bottomly et al. 2011). It is also available at http://bowtie-bio.sourceforge.net/recount/. GSE26024 dataset contains 21 samples, including two samples, 10 C57BL/6J (B6) and 11 DBA/2J (D2), and 36,536 genes (Korostynski et al. 2006; Wang et al. 2019b).

2.2 Methods for Identification of DEGs

For identifying the DEGs from the RNA-Seq dataset, several methods such as DESeq, DESeq2, EBSeq, edgeR, baySeq, limma, NBPSeq, etc., have been developed. In our study, three popular methods, DESeq2 (Love et al. 2014), edgeR (Robinson et al. 2010), and limma (Smyth et al. 2005), were used from Bioconductor (www.bioconductor.org) project to examine the DEGs between B6 and D2 samples. The following subsections explain a summary of these three methods.

2.3 DESeq2

DESeq2 is described based on the negative binomial distribution model (Love et al. 2014). A generalized linear model is used for DESeq2 and the model form is:

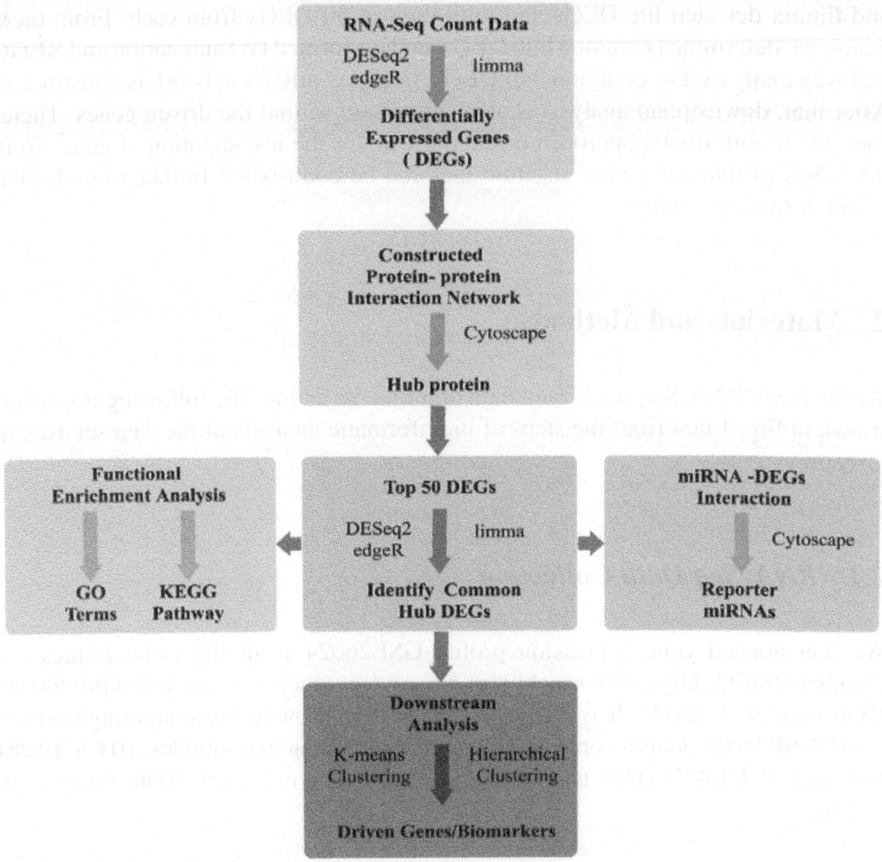

Fig. 1 RNA-Sequencing profiles of mouse striatum data analysis workflow (*Source* Created by the authors)

$$K_{ij} \sim NB(\mu_{ij,\alpha i})$$
$$\mu_{ij} = s_j q_{ij}$$
$$\log_2(q_{ij}) = x_j \beta_i$$

where, count K_{ij} is *i*-th gene and *j*-th sample model supported a negative binomial distribution; fitted mean and gene-specific dispersion parameters are denoted by μ_{ij} and α_i, respectively. The fitted mean is examined by a sample-specific size factor and a parameter, s_j and q_{ij}, respectively. The coefficients β_i calculated the $\log_2$-fold changes of the model matrix (**X**) each column for gene *i*. Sample and gene-dependent normalization factors s_{ij} are used after generalization of the model and the variance of counts K_{ij},

$$Var(K_{ij}) = [(K_{ij} - \mu_{ij})^2] = \mu_{ij} + \alpha_i \mu_{ij}^2$$

Maximum a posterior estimation of the $\log_2$-fold changes in β_i after incorporating a zero-centered normal prior provided by DESeq2 (Love et al. 2014).

2.4 *edgeR*

edgeR model and software was developed by Robinson et al. (2010). edgeR considered the hypothesis,

$$H_0 : \lambda_{j1} = \lambda_{j2} \text{ (Equally expressed)}$$

$$(and) H_A : \lambda_{j1} \neq \lambda_{j2} \text{ (Differentially expressed)}$$

In edgeR, the proportion of total reads, $\lambda_{jk(i)} = \sum_{i=1}^{C_k} \lambda_{ji}$, where, $\lambda_{jk(i)}$ is the jth genes of the kth group, and λ_{ji} is defined as the proportion of reads in the jth gene in an ith sample. Then the moderate mean $\mu_{ji} = n_i \lambda_{jk(i)}$, where, n_i is the ith library. According to the gene-wise or pair-wise assumption, the dispersion parameter ϕ is estimated by a maximizing conditional weighted log-likelihood,

$$WL(\phi_j) = l_j(\phi_j) + \alpha l_c(\phi_j)$$

where α is the weight, l_c is the maximum estimator denoted by $\hat{\phi}_j^{WL}$ which is considered as an empirical Bayesian solution. To estimate dispersion parameter, Robinson et al. 2010 proposed quantile-adjusted conditional maximum likelihood (qCML) and CML as follows,

$$y_{ji} \sim NB(\mu_{ji}, \phi)$$

The maximum likelihood estimator (MLE) becomes $\frac{\sum_{i \in c_j} y_{ji}}{\sum_{i \in c_j} n_i}$ and the dispersion parameter is given as $Z_j = \sum_{i=1}^{m_j} y_{ki}$ and the common likelihood function $l_c(\phi)$,

$$l_c(\phi) = \sum_{j=1}^{G} l_j(\phi) = \sum_{j=1}^{G} \sum_{k=1}^{K} [\sum_{g=1}^{m_k} \log\Gamma(y_{ki} + \phi^{-1}) +$$
$$\log\Gamma(n_k \phi^{-1}) - \log\Gamma(Z_k + n_k \phi^{-1}) - n_k \log\Gamma(\phi^{-1})]$$

To assess the perfect dispersion parameter ϕ, the common likelihood $l_c(\phi)$ is used and the MLE of $\lambda_{jk(i)}$ depending on ϕ. After estimating MLE, the hypothesis is tested, and the alternative hypothesis H_A is used for identifying differentially expressed genes.

2.5 *Limma*

Linear models for microarray data, *i.e.*, the limma tool is broadly used for the analysis of RNA-Seq data (Law et al. 2014). Different steps of limma for analyzing DEGs are described as follows:

(a) 1st step is the normalizing of the data. Suppose, data matrix r_{gi} defined the RNA-Seq read count, where row and column defined the genes and samples, respectively ($g = 1, 2 \ldots G; i = 1, 2, \ldots, nk$). Voom method is used to transform the read count data matrix to log-counts per million (log-cpm) as follows:

$$y_{gi} = \log\left(\frac{c_{gi} + 0.5}{C_i + 1} \times 10^6\right)$$

where C_i denotes the mapped reads for sample i,

$$C_i = \sum_{g=1}^{G} C_{gi}.$$

(b) 2nd step is the searching of low expression genes and filter them.
(c) 3rd step is the introduction of a linear model for analyzing DEGs that describes the treatment factors assigned to different RNA samples. The model used here is:

$$E(y_{gi}) = \mu_{gi} = x_i^T \beta_g.$$

Here, covariate vector x_i and an unknown coefficient β_g represent $\log_2$-fold changes with the range of conditions of the experiment. In matrix form,

$$E(y_g) = X\beta_g$$

Here, a log-cpm value of vector is y_g for gene g and the design matrix is X with column x_i. The fitted model is

$$\hat{\mu}_{gi} = x_i^T \hat{\beta}_g$$

The mean log-cpm is transformed to mean log count value by:

$$\tilde{c} = \bar{y}_g + \log_2(\tilde{C}) - \log_2(10^6)$$

Here, the geometric mean is $\tilde{C}$. The log-cpm fitted values $\hat{\mu}_{gi}$ are transformed into fitted counts by

$$\hat{\lambda}_{gi} = \hat{\mu}_{gi} + \log_2(C_i + 1) - \log_2(10^6)$$

(d) Calculated voom weights using LOWESS curve (Cleveland, 1979) that is statistically robust and used to describe a piecewise function lo () which is linear. After that, the voom weight is $w_{gi} = lo(\hat{\lambda}_{gi})^{-4}$ called voom precision.
(e) Then fitted the contrasts of coefficient. The contrast is given by $\beta_g = M^T \alpha_g$, where, M defined the contrasts matrix, $\hat{\beta}_{gi} | \beta_{gi}, \sigma_g^2 \sim N(\beta_{gi}, v_{gi}\sigma_g^2)$.
(f) Empirical Bayes is used for getting better estimates, and it assumes the inverse of Chi-square prior σ_g^2 with mean s_0^2, f_0 is the degrees of freedom, and f_g is the residual degree. The posterior values for the residual variances are

$$\tilde{s}_g^2 = \frac{f_0 s_0^2 + f_g s_g^2}{f_0 + f_g}$$

Then the moderate t-statistic is

$$\tilde{t}_{gi} = \frac{\hat{\beta}_{gi}}{u_{gi}\tilde{s}_g}$$

(g) Adjust p-values for false discovery rate, and access the results that make sense for identifying differentially expressed genes.

2.6 *Methods for Functional Analysis of DEGs*

Functional analysis is carried out for annotations of DEGs and to explain their biological insights.

2.7 *PPI Analysis of DEGs*

PPI network represents the interaction of proteins, where nodes and edges represent the proteins and their interaction. Search tool for the Retrieval of Interacting Genes (STRING) database (http://www.string-db.org/) was used to collect information for DEGs (Szklarczyk et al. 2015), and an interaction network was considered where combined score > 0.4. Cytoscape software version 3.7.1 was used to visualize the regulatory network of their corresponding genes (Su et al. 2014). For the analysis of core genes, Network Analyzer in Cytoscape software was used for the interaction network.

2.8 GO Enrichment and KEGG Pathway Analysis of DEGs

Normally, high-throughput genomics or transcriptomics data is annotated by the GO enrichment analysis (Ashburner et al. 2000). Additionally, KEGG is a knowledge-based database used to manage natural pathways and infections. A significant genes list was submitted to the Gene Ontology (http://www.geneontology.org/) and KEGG pathway (http://www.genome.jp/kegg/) for inspecting over-represented GO and pathway classes. GO is studied to predict the possible elements of the DEGs in BP, biological process or GO process; MF, molecular function or GO function; and CC, cellular component or GO component. KEGG pathway analysis is performed for gene functions investigation (Altermann and Klaenhammer 2005), connecting genomic information with higher-level systemic functions, etc. In addition, we have considered statistically significant over-represented pathway categories in KEGG pathway enrichment analysis.

2.9 miRNAs-Target Gene Interactions of DEGs

miRNAs molecules are involved with numerous physiological and disease processes. Each miRNA is assumed to control manifold genes to select probable miRNA–mRNA interaction within the hub genes network (Lim et al. 2003). We used miRDB (http://mirdb.org/) for miRNAs-target gene interactions (Wong and Wang 2015). Cytoscape software was used to develop a regulatory miRNA–mRNA network.

2.10 Downstream Analysis of DEGs

Clustering is crucial for understanding gene expression data. Clusters are obtained by the similarity of genes in a gene expression profile. The popular k-means clustering algorithm is used for clustering DEGs. We also used hierarchical clustering that is also known as hierarchical cluster analysis. It attempts to group genes into small clusters and to group clusters into higher-level systems (Eisen et al. 1998; Kuklin et al. 2001). A common method for visualization of gene expression data using hierarchical clustering is the *heatmap*. The *heatmap* may also be combined with hierarchical clustering methods, which may split genes into groups and/or samples together, and support to display DEGs expression pattern. This may also be useful for identifying genes that are commonly regulated, or biological signatures related to a selected condition.

Table 1 Number of DEGs with p-value < 0.01

Methods	DEGs	Up-regulated DEGs	Down-regulated DEGs
DESeq2	736	37	699
edgeR	757	48	709
limma	530	31	499

Source Created by the authors

3 Results

3.1 Identified DEGs

DESeq2, edgeR, and limma methods identified DEGs summarized in Table 1. We identified DEGs by considering p-value < 0.01 and discriminate up-regulated and down-regulated genes based on the cut-off criteria, log FC > 2.0 and log FC < −2.0, respectively.

3.2 PPI Analysis of DEGs

According to the information in the STRING database, the gene interaction network contained many nodes and edges. Nodes and edges are listed in Table 2. DEGs are demonstrated by the nodes, and interactions between DEGs are showed by the edges. Predicted scores (degree) are used to rank core genes.

We selected the top 50 high-degree hub DEGs for each method, and the distribution of the top 50 DEGs in the interaction network is shown in Fig. 2. The relationship between the data points and comparing points on the line are roughly 0.821, 0.844, and 0.842, and the R^2 values are 0.912, 0.907, and 0.897 for DESeq2, edgeR, and limma, respectively.

Venn diagram discovered 21 common hub DEGs among the top 50 high-degree hub DEGs as shown in Fig. 3. These 21 DEGs are Bdkrb2, C5ar1, C3ar1, Fpr1, Ccr6, Ptgs2, Mki67, Tas1r2, Sstr5, Ccl28, Aplnr, Apln, Gpr55, B2m, H2-K1, F2r, Dnajc3, Trhr, Polr1a, Adcy4 and Mog. Venn diagram is drawn using the R package "VennDiagram." Again the interaction network of the 21 common hub DEGs is made

Table 2 Nodes and edges were identified based on p-value < 0.01

Methods	Nodes	Edges
DESeq2	725	1441
edgeR	744	1713
limma	520	678

Source Created by the authors

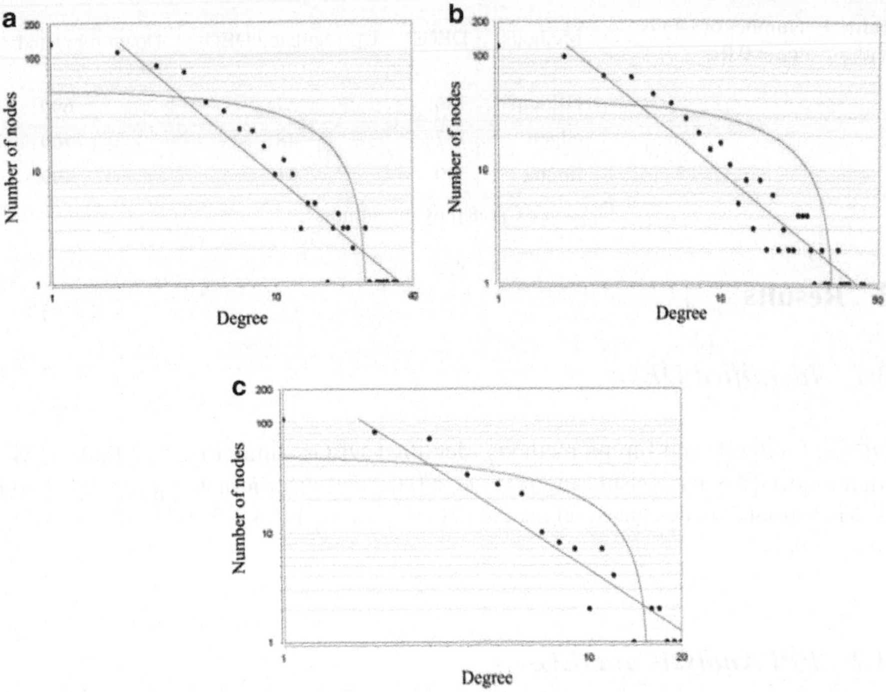

Fig. 2 Nodes-degree relationship where (**a**) DEGs found through DEseq2, (**b**) DEGs found through edgeR, and (**c**) DEGs found through limma. The dot (black) node indicates the core genes, the curve (red) indicates the fitted line, and the straight (blue) line indicates the power law. (*Source* Created by the authors)

Fig. 3 Venn diagram of the DEGs detected by the three methods (*Source* Created by the authors)

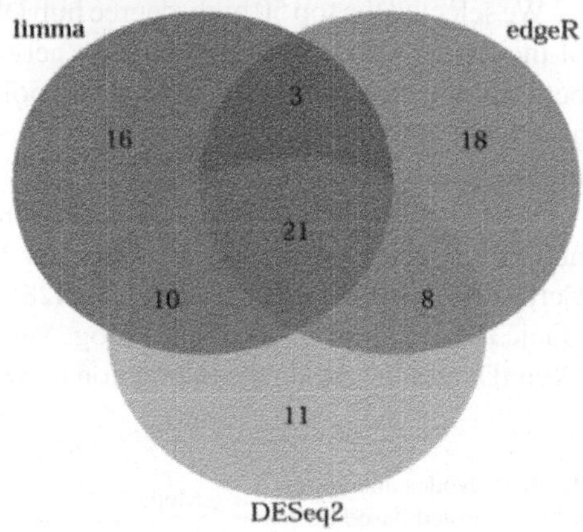

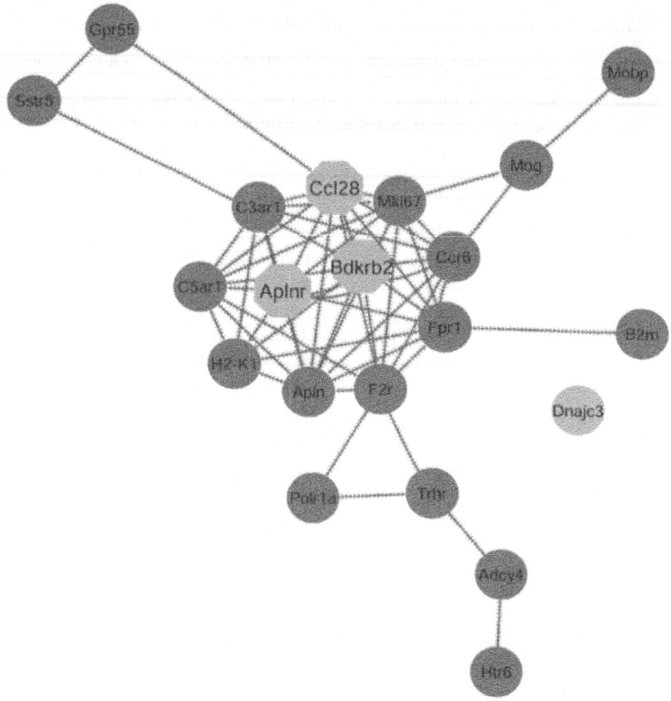

Fig. 4 The interaction network of the common 21 hub DEGs. The hub genes are indicated by the nodes, and the interactions between the hub genes are indicated by the edges. (*Source* Created by the authors)

by the STRING database containing 21 nodes and 67 edges (Fig. 4). Up-regulated hub genes Bdkrb2, Aplnr, and Ccl28 were highlighted by a different color from other down-regulated genes.

3.3 GO Enrichment Analysis of DEGs

Functional analysis of the common 21 hub DEGs is clarified through GO analysis. GO function indicates that these 21 hub DEGs are enriched in G-protein coupled peptide receptor activity, peptide binding, signaling receptor binding, etc. GO process indicates that these 21 hub DEGs are enriched in cell death, response to stimulus, signaling, homeostatic process, immune system process, response to stimulus, blood vessel development, cAMP-mediated signaling, heart development, and other biological processes. For the GO component, the 21 hub DEGs were enriched in the plasma membrane, an integral component of the plasma membrane, cytoplasmic vesicle, and so on. GO analysis results of these DEGs are explained in Table 3.

Table 3 GO enrichment analysis of common 21 hub DEGs

Gene	Gene title	GO: function	GO: process	GO: component
Bdkrb2	Bradykinin receptor, beta 2	G-protein coupled peptide receptor activity, peptide binding	Cell death, response to stimulus, signaling	Plasma membrane
C5ar1	C5a anaphylatoxin chemotactic receptor 1	G-protein coupled receptor activity, phospholipase C activity	Activation of phospholipase C activity, immune response, immune response-activating cell surface receptor signaling pathway, inflammatory response	Intracellular, cytoplasmic vesicle
C3ar1	C3a anaphylatoxin chemotactic receptor	G-protein coupled receptor activity, phospholipase C activity	Activation of phospholipase C activity, immune response, immune response-activating cell surface receptor signaling pathway, inflammatory response, inositol phosphate-mediated signaling	Intracellular, plasma membrane
Fpr1	fMet-Leu-Phe receptor	G-protein coupled peptide receptor activity, phospholipase C activity	Activation of phospholipase C activity, immune response, immune response-activating cell surface receptor signaling pathway, inflammatory response, inositol phosphate-mediated signaling	Intracellular, plasma membrane
Ccr6	C–C chemokine receptor type 6	G-protein coupled peptide receptor activity	Calcium-mediated signaling, cell chemotaxis, immune response, positive regulation of cytosolic calcium ion concentration	External side of the plasma membrane, intracellular
Ptgs2	Prostaglandin G/H synthase 2	Oxidoreductase activity, cell death, response to stimulus	Immune system process, system development, cell differentiation	Endoplasmic reticulum, plasma membrane

(continued)

Table 3 (continued)

Gene	Gene title	GO: function	GO: process	GO: component
Mki67	Proliferation marker protein Ki-67	DNA binding	Cell population proliferation, system development	Non-membrane-bounded organelle, nucleus
Tas1r2	Taste receptor type 1 member 2	G-protein coupled receptor activity, taste receptor activity	Sensory perception of sweet taste response to stimulus, signaling	Integral component of plasma membrane
Sstr5	Somatostatin receptor type 5	G-protein coupled receptor activity, neuropeptide binding	Neuropeptide signaling pathway, response to stimulus, signaling, homeostatic process	Integral component of plasma membrane
Ccl28	C–C motif chemokine 28	Signaling receptor binding	Homeostatic process, immune system process, response to stimulus	Cytoplasmic vesicle
Aplnr	Apelin receptor	G-protein coupled peptide receptor activity	Blood vessel development, cAMP-mediated signaling, heart development	Intracellular, plasma membrane
Apln	Apelin	Signaling receptor binding, extracellular region	Cell population proliferation, establishment of localization, signaling	Extracellular region, signaling receptor binding
Gpr55	G-protein coupled receptor 55	G-protein coupled receptor activity, phospholipase C activity	Rho protein signal transduction activation of phospholipase C activity, inositol phosphate-mediated signaling, positive regulation of cytosolic calcium ion concentration	Integral component of plasma membrane intracellular
B2m	Beta-2-microglobulin		Homeostatic process, cell differentiation, system development, immune system process	Cytosol, Golgi apparatus, plasma membrane

(continued)

Table 3 (continued)

Gene	Gene title	GO: function	GO: process	GO: component
H2-K1	H-2 class I histocompatibility antigen	Peptide binding signaling receptor binding	Adaptive immune response, immune effector process, positive regulation of adaptive immune response	
F2r	Proteinase-activated receptor 1	G-protein coupled receptor activity, phospholipase C activity	Rho protein signal transduction, activation of phospholipase C activity, inositol phosphate-mediated signaling, positive regulation of cytosolic calcium ion concentration	Integral component of plasma membrane, intracellular
Dnajc3	DnaJ homolog subfamily C member 3	Chaperone binding, unfolded protein binding, signaling receptor binding	Protein folding in endoplasmic reticulum, cell differentiation, cellular component organization, system development, immune system process	Endoplasmic reticulum, plasma membrane, Golgi apparatus
Trhr	Thyrotropin-releasing hormone receptor	Signaling receptor activity	Muscle contraction, sensory perception, homeostatic process, response to stimulus, signaling	Plasma membrane
Polr1a	DNA-directed RNA polymerase subunit RPA1	RNA polymerase I activity, transferase	Nucleic acid-templated transcription	DNA-directed RNA polymerase I complex, nuclear chromatin
Adcy4	Adenylate cyclase type 4	G-protein coupled receptor activity, adenylate cyclase activity	Activation of adenylate cyclase activity, adenylate cyclase-activating G-protein coupled receptor signaling pathway, regulation of adenylate cyclase activity	Integral component of plasma membrane, intracellular

(continued)

Table 3 (continued)

Gene	Gene title	GO: function	GO: process	GO: component
Mog	Myelin-oligodendrocyte glycoprotein	Signaling receptor binding, carbohydrate derivative binding	T cell receptor signaling pathway, immune response, immune system process, response to stimulus	External side of plasma membrane, leaflet of membrane bilayer

Source Created by the authors

3.4 KEGG Pathway Analysis of DEGs

In the analysis of the KEGG pathway, we have considered a false discovery rate (FDR) less than 0.05 and found out significant genes. KEGG pathway analysis exposed and targeted pathways enriched in neuroactive ligand–receptor interaction, pathways in cancer, ovarian steroidogenesis, and other significant pathways described in Table 4.

Pathway ranking associated with genes is displayed in Fig. 5. The first-ranked staphylococcus aureus infection pathway had the 6% genes that involved C5ar1, C3ar1, and Fpr1. The second is the complement and coagulation cascades pathway with 4.5% related genes that are Bdkrb2, C5ar1, C3ar1, and F2r. The third, regulation of lipolysis in adipocytes pathway, had the 3.65% related genes that included Ptgs2, Adcy4. The fourth, the ovarian steroidogenesis pathway, had 3.51% related genes that are Adcy4, Ptgs2. And, the fifth, neuroactive ligand–receptor interaction pathway, had the 3% related genes that are Bdkrb2, C5ar1, C3ar1, Fpr1, Sstr5, Aplnr, Apln, and F2r.

3.5 miRNA–mRNA Network Construction for DEGs

The common 21 hub DEGs were closely associated with related miRNA and predicted potential miRNAs. The prediction scores were likewise gathered from the miRDB database, and therefore the miRNA–mRNA with a high score implied near-potential function of miRNA inside the guideline of the objective mRNA. The miRNA–mRNA network appeared in Fig. 6 with cutoff > 70.

3.6 Downstream Analysis for DEGs

Cluster analysis of 21 hub DEGs is shown in Fig. 7. Two popular clustering methods, hierarchical clustering and k-means, were applied for finding the similarity of DEGs. We divided DEGs into three clusters for both methods. In the k-means algorithm, it observed that Ptgs2, Mog, and Dnajc3 are clustered together in Group 1; Polr1a,

Table 4 KEGG pathway analysis of common 21 hub DEGs

Pathway	Description	Genes count	Associated genes	FDR
mmu04080	Neuroactive ligand–receptor interaction	8 of 284	Bdkrb2, C5ar1, C3ar1, Fpr1, Sstr5, Aplnr, Apln, F2r	1.3E−08
mmu04610	Complement and coagulation cascades	4 of 88	Bdkrb2, C5ar1, C3ar1, F2r	6.9E−05
mmu05150	Staphylococcus aureus infection	3 of 50	C5ar1, C3ar1, Fpr1	0.0005
mmu04020	Calcium signaling pathway	4 of 180	Bdkrb2, F2r, Trhr, Adcy4	0.0005
mmu04371	Apelin signaling pathway	3 of 134	Aplnr, Apln, H2-K1	0.0050
mmu04062	Chemokine signaling pathway	3 of 179	Ccr6, Ccl28, Adcy4	0.0095
mmu04024	cAMP signaling pathway	3 of 194	Sstr5, F2r, Adcy4	0.0103
mmu04015	Rap1 signaling pathway	3 of 207	Fpr1, F2r, Adcy4	0.0108
mmu05200	Pathways in cancer	4 of 522	Adcy4, F2r, Ptgs2, Bdkrb2	0.0129
mmu04923	Regulation of lipolysis in adipocytes	2 of 55	Ptgs2, Adcy4	0.0129
mmu04913	Ovarian steroidogenesis	2 of 57	Adcy4, Ptgs2	0.0129
mmu04612	Antigen processing and presentation	2 of 78	B2m, H2-K1	0.0189
mmu04742	Taste transduction	2 of 86	Tas1r2, Adcy4	0.0210
mmu04750	Inflammatory mediator regulation of TRP channels	2 of 119	Adcy4, F2r	0.0363
mmu04611	Platelet activation	2 of 122	F2r, Adcy4	0.0363
mmu04921	Oxytocin signaling pathway	2 of 149	Ptgs2, Adcy4	0.0463
mmu04723	Retrograde endocannabinoid signaling	2 of 145	Ptgs2, Adcy4	0.0463
mmu04072	Phospholipase D signaling pathway	2 of 145	Adcy4, F2r	0.0463
mmu04022	cGMP-PKG signaling pathway	2 of 164	Bdkrb2, Adcy4	0.0492
mmu04141	Protein processing in endoplasmic reticulum	1 of 164	Dnajc3	0.0496

Source Created by the authors

Apln, and B2m belong to Group 3; and remaining DEGs are contained in Group 2. Hierarchical clustering using heatmap presentation of DEGs observed that Ptgs2, Mog, Dnajc3, Apln, and B2m are clustered together in Group 1; Polr1a, Htr6, F2r, Sstr5, and Trhr belong to Group 3; and the remaining DEGs are clustered together in Group 2.

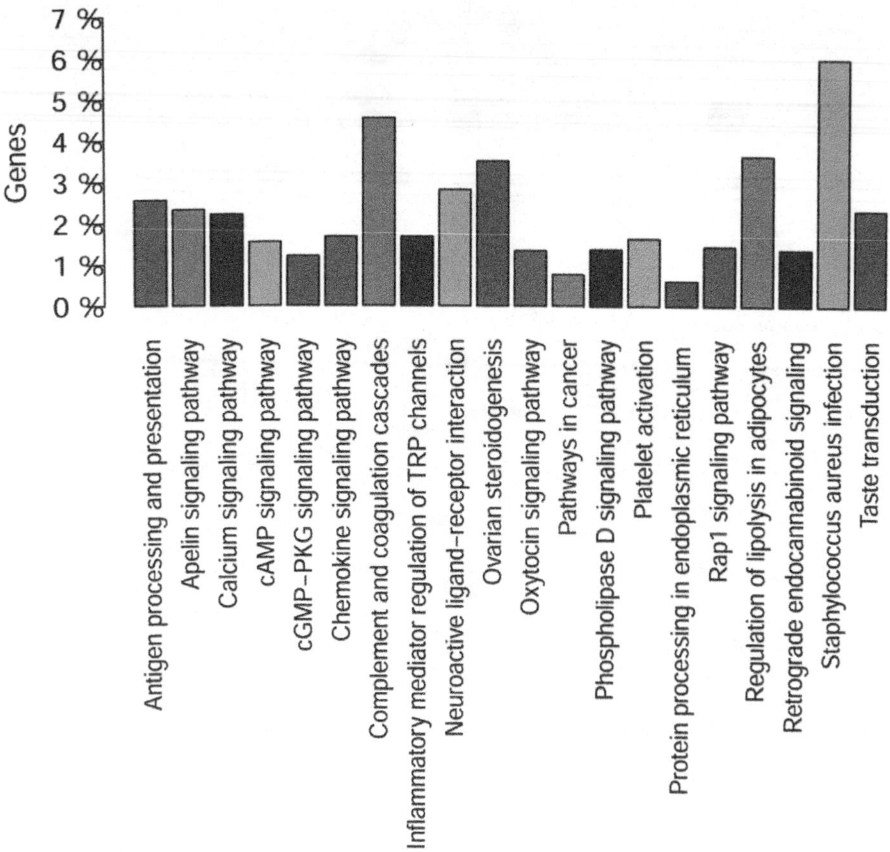

Fig. 5 KEGG analysis of common 21 hub DEGs. The different color means different pathways. (*Source* Created by the authors)

4 Discussion

The most recognized mouse strains such as C57BL/6J (B6) and DBA/2J (D2) samples are widely used in neuroscience research (Sandberg et al. 2000). In the current study, the mouse striatum gene expression profile of GSE26024 was downloaded, and to identify core genes bioinformatic analysis was performed. These investigations confirmed that 736, 757, and 530 DEGs are identified using DESeq2, edgeR, and limma with 37, 48, and 31 up-regulated genes, respectively (Table 1). Furthermore, protein–protein interaction network analysis, GO, KEGG pathway, construction of miRNA–mRNA network, and downstream analysis were executed to access the biomarkers or the core genes.

Table 2 displayed the nodes and edges of the DEGs assessed by the three different methods. The protein–protein interaction network investigation recognized the top 50 highest-degree hub genes of DEGs selected from each DEGs set. Figure 2 explained

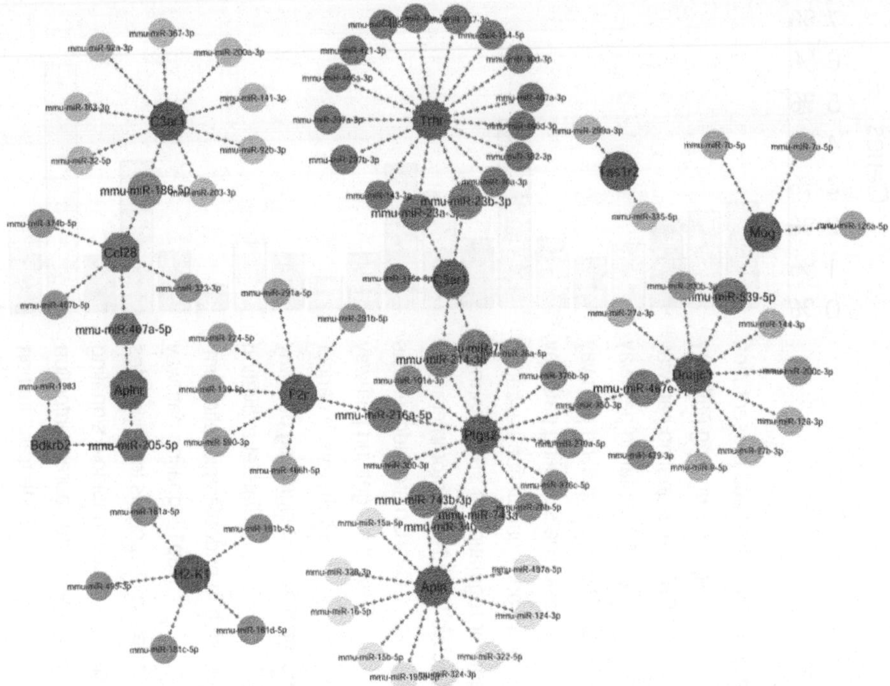

Fig. 6. miRNA–mRNA interaction network of DEGs (*Source* Created by the authors)

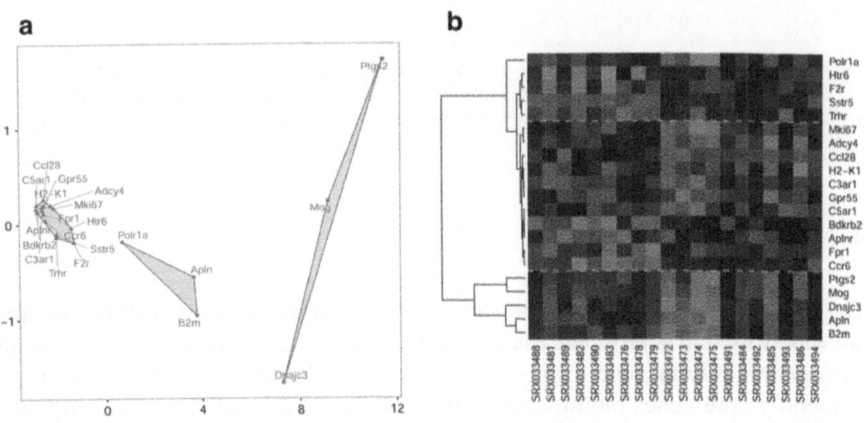

Fig. 7 Cluster analysis of common 21 hub DEGs. (**a**) K-means clustering and (**b**) Hierarchical clustering of DEGs with three clusters (*Source* Created by the authors)

nodes-degree relationship of the top 50 hub DEGs. It describes core gene distribution by giving generally high certainty that the basic model is linear in the interaction network. Figure 3 is a Venn diagram displaying and identifying 21 common hub genes based on the top 50 hub DEGs. Among the common 21 hub DEGs, DESeq2 and limma found Bdkrb2 and Ccl28, and edgeR found Aplnr up-regulated genes. The other genes are observed to be down-regulated. We also performed gene interaction network analysis for these 21 common hub genes and observed that the gene "Dnajc3" had no interaction with other genes (Fig. 4).

To disclose the underlying molecular mechanisms, we have characterized the possible GO terms and biological pathways of common hub genes. GO enrichment analysis is displayed in Table 3. The up-regulated DEGs, Bdkrb2 and Aplnr, are mainly involved in the same functional terms such as plasma membrane, and Ccl28 is associated with cytoplasmic vesicle; contrariwise, down-regulated DEGs are observed to be rich in biological intracellular, plasma membrane, endo plasmic reticulum, DNA-directed RNA polymerase I complex, non-membrane-bounded organelle, and so on.

Besides, KEGG pathway analysis is used for identifying the functional analysis of DEGs. According to KEGG pathway analysis, multiple genes are associated with same pathway as well as a same gene is associated with several pathways. Bdkrb2 is enriched in the Neuroactive ligand–receptor interaction pathway, Complement and coagulation cascades, Calcium signaling pathway, and Pathways in cancer. C5ar1 and C3ar1 regulate the Neuroactive ligand–receptor interaction, Complement and coagulation cascades, and Staphylococcus aureus infection. Fpr1 is associated with Neuroactive ligand–receptor interaction, Staphylococcus aureus infection, and Rap1 signaling pathway. Adcy4 is associated with several pathways such as Calcium signaling pathway, Chemokine signaling pathway, cAMP signaling pathway, Rap1 signaling pathway, Pathways in cancer, Regulation of lipolysis in adipocytes, Ovarian steroidogenesis, Taste transduction, Inflammatory mediator regulation of TRP channels, Platelet activation, Oxytocin signaling pathway, Retrograde endocannabinoid signaling and Phospholipase D signaling pathway, and so on (Table 4). The up-regulated DEGs, Bdkrb2 and Aplnr, are significantly enriched in Neuroactive ligand-receptor interaction pathway, while Ccl28 is enriched in Chemokine signaling pathway. Figure 5 describes the percentage of genes which are involved with different pathways.

We also have constructed miRNA–mRNA network for the common hub genes (Fig. 6). Multiple hub genes are observed to be connected with miRNAs. Trhr and C5ar1 hub genes related to mmu-miR-23a-3p and mmu-miR-23b-3p. MiR-23a downregulation is the following experiment of traumatic brain injury (Sabirzhanov et al. 2014) and MiR-23b is involved in cancer aggressive (Grossi et al. 2018). Ptgs2 and C5ar1 genes are connected with mmu-miR-761 and mmu-miR-214-3p. MiR-761 is involved in suppressing the remodeling of nasal mucosa (Cheng et al. 2020). F2r and Ptgs2 are observed to be connected with mmu-miR-216a-5p while Dnajc3 and Ptgs2 are connected with mmu-miR-467e-5p, Dnajc3 and Mog are connected with mmu-miR-539-5p, Apln and Ptgs2 are connected with mmu-miR-743a-3p, mmu-miR-743b-3p, mmu-miR-340-5p, and C3ar1and Ccl28 are connected with mmu-miR-186-5p. It is more interesting that up-regulated hub genes, Ccl28 and Aplnr, are associated with mmu-miR-467a-5p while Aplnr and Bdkrb2 are interconnected with mmu-miR-205-5p. MiR-467a is highly expressed in tumor suppressors (Inoue et al. 2017) and MiR-205 upregulation determines the aggressiveness and metastatic activity of malignant tumors (Dahmke et al. 2013).

The downstream analysis (Fig. 7) explained the cluster of 21 hub DEGs, in which maximum DEGs clustered in group 2 and a small number of DEGs clustered in group 1 and 3. We observed that the up-regulated DEGs, Ccl28, Aplnr, and Bdkrb2, belong to the same cluster (group 2) of both k-means and hierarchical clustering methods. From the above discussions, we may highlight that the genes Ccl28, Aplnr, and Bdkrb2 are crucial genes and might be the driven genes. More importantly, they might be the biomarkers for further neuroscience research.

5 Conclusions

In summary, DEGs are identified from RNA-Seq profiles of mouse striatum using the three popular DEGs calculation methods, and applied PPI network on DEGs. Then, the 21 common hub DEGs were recognized including the up-regulated genes Bdkrb2, Aplnr, and Ccl28. Analysis of GO and KEGG pathway identified significant genes to explore the biological insights of the DEGs. The downstream analysis explained that Bdkrb2, Aplnr, and Ccl28 genes belong to the same group. Finally, we have concluded that the hub genes, Bdkrb2, Aplnr, and Ccl28, might be the driven genes in inbred mouse strains. These identified driven genes might be promising candidates or biomarkers for further neuroscience research. Furthermore, experimental validation is needed and should be made in future studies.

Acknowledgements We are thankful to the reviewers for their many valuable suggestions to improve the manuscript.

References

Alioto, T., Behr, J., Bohnert, R., Campagna, D., Davis, C. A., Dobin, A., et al. (2013). Systematic evaluation of spliced alignment programs for RNA-seq data. *Nature Methods, 10*, 1185–1191.

Altermann, E., & Klaenhammer, T. R. (2005). PathwayVoyager: Pathway mapping using the Kyoto Encyclopedia of Genes and Genomes (KEGG) database. *BMC Genomics, 6*, 60.

Anders, S., McCarthy, D. J., Chen, Y., Okoniewski, M., Smyth, G. K., Huber, W., et al. (2013). Count-based differential expression analysis of RNA sequencing data using R and Bioconductor. *Nature Protocols, 8*, 1765.

Ashburner, M., Ball, C. A., Blake, J. A., Botstein, D., Butler, H., Cherry, J. M., et al. (2000). Gene ontology: Tool for the unification of biology. *Nature Genetics, 25*, 25–29.

Bacher, R., & Kendziorski, C. (2016). Design and computational analysis of single-cell RNA-sequencing experiments. *Genome Biology, 17*, 63.

Bottomly, D., Walter, N. A., Hunter, J. E., Darakjian, P., Kawane, S., Buck, K. J., et al. (2011). Evaluating gene expression in C57BL/6J and DBA/2J mouse striatum using RNA-Seq and microarrays. *PloS One*, *6*(3), e17820.

Cheng, J., Chen, J., Zhao, Y., Yang, J., Xue, K., & Wang, Z. (2020). MicroRNA-761 suppresses remodeling of nasal mucosa and epithelial–mesenchymal transition in mice with chronic rhinosinusitis through LCN2. *Stem Cell Research and Therapy, 11*, 1–11.

Cleveland, W. S. (1979). Robust locally weighted regression and smoothing scatterplots. *Journal of the American Statistical Association, 74*, 829–836.

Dahmke, I. N., Backes, C., Rudzitis-Auth, J., Laschke, M. W., Leidinger, P., Menger, M. D., et al. (2013). Curcumin intake affects miRNA signature in murine melanoma with mmu-miR-205-5p most significantly altered. *PLoS One, 8*, e81122.

Eisen, M. B., Spellman, P. T., Brown, P. O., & Botstein, D. (1998). Cluster analysis and display of genome-wide expression patterns. *Proceedings of the National Academy of Sciences. National Acad Sciences, 95*, 14863–14868.

Grossi, I., Salvi, A., Baiocchi, G., Portolani, N., & De Petro, G. (2018). Functional role of microRNA-23b-3p in cancer biology. *MicroRNA, 7*, 156–166.

Hovatta, I., Tennant, R. S., Helton, R., Marr, R. A., Singer, O., Redwine, J. M., et al. (2005). Glyoxalase 1 and glutathione reductase 1 regulate anxiety in mice. *Nature, 438*, 662–666.

Huber, W., Carey, V. J., Gentleman, R., Anders, S., Carlson, M., Carvalho, B. S., et al. (2015). Orchestrating high-throughput genomic analysis with Bioconductor. *Nature Methods*. Nature Publishing Group, *12*, 115.

Hwang, B., Lee, J. H., & Bang, D. (2018). Single-cell RNA sequencing technologies and bioinformatics pipelines. *Experimental and Molecular Medicine, 50*, 1–14.

Inoue, K., Hirose, M., Inoue, H., Hatanaka, Y., Honda, A., Hasegawa, A., et al. (2017). The rodent-specific microRNA cluster within the Sfmbt2 gene is imprinted and essential for placental development. *Cell Reports, 19*, 949–956.

Korostynski, M., Kaminska-Chowaniec, D., Piechota, M., & Przewlocki, R. (2006). Gene expression profiling in the striatum of inbred mouse strains with distinct opioid-related phenotypes. *BMC Genomics, 7*, 146.

Kuklin, A., Shah, S., Hoff, B., & Shams, S. (2001). *Data management in microarray fabrication, image processing, and data mining* (p. 115). Technologies and Experimental Strategies. CRC Press.

Law, C. W., Chen, Y., Shi, W., & Smyth, G. K. (2014). voom: Precision weights unlock linear model analysis tools for RNA-seq read counts. *Genome Biology, 15*, R29.

Lim, L. P., Glasner, M. E., Yekta, S., Burge, C. B., & Bartel, D. P. (2003). Vertebrate microRNA genes. *Science. American Association for the Advancement of Science, 299*, 1540–1540.

Love, M. I., Huber, W., & Anders, S. (2014). Moderated estimation of fold change and dispersion for RNA-seq data with DESeq2. *Genome Biology, 15*, 550.

Pop, M., & Salzberg, S. L. (2008). Bioinformatics challenges of new sequencing technology. *Trends in Genetics, 24*, 142–149.

Robinson, M. D., McCarthy, D. J., & Smyth, G. K. (2010). edgeR: A Bioconductor package for differential expression analysis of digital gene expression data. *Bioinformatics, 26*, 139–140.

Sabirzhanov, B., Zhao, Z., Stoica, B. A., Loane, D. J., Wu, J., Borroto, C., et al. (2014). Down-regulation of miR-23a and miR-27a following experimental traumatic brain injury induces neuronal cell death through activation of proapoptotic Bcl-2 proteins. *Journal of Neuroscience, 34*, 10055–10071.

Sandberg, R., Yasuda, R., Pankratz, D. G., Carter, T. A., Del Rio, J. A., Wodicka, L., et al. (2000). Regional and strain-specific gene expression mapping in the adult mouse brain. *Proceedings of the National Academy of Sciences, 97*, 11038–11043.

Shendure, J., & Ji, H. (2008). Next-generation DNA sequencing. *Nature Biotechnology, 26*, 1135.

Smyth, G. K., Ritchie, M., Thorne, N., & Wettenhall, J. (2005). LIMMA: Linear models for microarray data. Bioinformatics and computational biology solutions using R and bioconductor. *Statistics for Biology and Health*, 397–420.

Soneson, C., & Delorenzi, M. (2013). A comparison of methods for differential expression analysis of RNA-seq data. *BMC Bioinformatics, 14*, 91.

Su, G., Morris, J. H., Demchak, B., & Bader, G. D. (2014). Biological network exploration with Cytoscape 3. *Current Protocols in Bioinformatics, 47*, 8–13.

Szklarczyk, D., Franceschini, A., Wyder, S., Forslund, K., Heller, D., Huerta-Cepas, J., et al. (2015). STRING v10: Protein–protein interaction networks, integrated over the tree of life. *Nucleic Acids Research, 43*, D447–D452.

Van Dijk, E. L., Auger, H., Jaszczyszyn, Y., & Thermes, C. (2014). Ten years of next-generation sequencing technology. *Trends in Genetics., 30*, 418–426.

Wang, T., Li, B., Nelson, C. E., & Nabavi, S. (2019a). Comparative analysis of differential gene expression analysis tools for single-cell RNA sequencing data. *BMC Bioinformatics, 20*, 40.

Wang, J., Geisert, E. E., & Struebing, F. L. (2019b). RNA sequencing profiling of the retina in C57BL/6J and DBA/2J mice: Enhancing the retinal microarray data sets from GeneNetwork. *Molecular Vision, 25*, 345.

Wong, N., & Wang, X. (2015). miRDB: An online resource for microRNA target prediction and functional annotations. *Nucleic Acids Research, 43*, D146–D152.

Role of Serum High-Sensitivity C-Reactive Protein Level as Risk Factor in the Prediction of Coronary Artery Disease in Hyperglycemic Subjects

Md. Saiful Islam, Rowshanul Habib, Md. Rezaul Karim, and Tanzima Yeasmin

Abstract In this study, we have evaluated the clinical value of the high-sensitivity C-reactive protein (hs-CRP) level in predicting the risk of coronary artery disease (CAD) in hyperglycemic subjects of Bangladesh. A total of 201 participants were selected for this study and fasting venous blood samples are collected from them for measuring fasting plasma glucose (FPG), serum total cholesterol (TC), triglycerides (TG), LDL-cholesterol (LDL-C) and HDL-cholesterol (HDL-C), hs-CRP, apolipoprotein A-1, apolipoprotein B and lipoprotein(a). CAD risk of study subjects was estimated with the use of Framingham Risk Score (FRS). Out of 201, 91 study participants are found as normal fasting glucose (NFG), 56 as impaired fasting glucose (IFG) and 54 as diabetes mellitus. The average levels of TC, TG, LDL-C, HDL-C, TC/HDL-C ratio, apolipoprotein A-1, apolipoprotein B and lipoprotein(a) in NFG, IFG and diabetes are not significantly different between each group. Statistically significant ($p < 0.001$) differences are observed between each group of hs-CRP levels. Among the components of FRS; age, systolic blood pressure and HDL-C are significantly correlated with an increase in concentration of FPG. The estimated Framingham 10-year risk of CAD and hs-CRP levels are significantly increasing with the concentration of FPG. Before and after adjusting for covariates (age, sex, smoking status, TC and HDL-C), it is found that FPG is significantly associated with hs-CRP level. Interestingly, serum hs-CRP levels are significantly increased in the higher FPG groups before and after adjustment by covariates. Finally, this study demonstrates hs-CRP as a stronger predictor of cardiovascular events in hyperglycemic subjects thereby helping to assess the risk of CAD induced by hyperglycemia.

Md. S. Islam · R. Habib · T. Yeasmin (✉)
Department of Biochemistry and Molecular Biology, University of Rajshahi, Rajshahi-6205, Bangladesh
e-mail: yeasmin_bio@ru.ac.bd

R. Habib
e-mail: mrhabib@ru.ac.bd

Md. R. Karim
Department of Applied Nutrition and Food Technology, Islamic University, Kushtia-7003, Bangladesh

B. K. Sinha et al. (eds.), *Data Science and SDGs*,
https://doi.org/10.1007/978-981-16-1919-9_10

Keywords Hyperglycemia · Cardiovascular disease · High-sensitivity C-reactive protein · Framingham risk score

1 Introduction

Hyperglycemia found under diabetic conditions is the most prevalent and serious metabolic disease all over the world, and it facilitates the progression of atherosclerosis and subsequent cardiovascular disease (CVD), which, in turn, is the leading cause of mortality and morbidity in diabetic patients (Nathan 1993; Steiner 1985). People with diabetes and IGT (impaired glucose tolerance) are at two- to four-fold greater risk of developing CVD than people without diabetes (Laakso and Kuusisto 1996; Ruderman et al. 1992). CVD has a multifactorial etiology with several potentially modifiable risk factors among which the two most important modifiable cardiovascular risk factors are smoking and abnormal lipids. Hypertension, diabetes, psychosocial factors and abdominal obesity are the next most important but their relative effects vary in different regions of the world (Yusuf et al. 2004).

The Framingham Risk Score (FRS) is an important global clinical tool to enable clinicians to estimate cardiovascular risk prediction (Anderson et al. 1991; Wilson et al. 1998). A large systematic review of cardiovascular risk assessment in primary prevention has shown that the performance of Framingham risk scores varies considerably between populations and that accuracy relates to the background risk of the population to which it has been applied (Brindle et al. 2006; Koenig et al. 2004). The accuracy of Framingham cardiovascular risk assessments is limited by the exclusion of certain risk factors including obesity, physical inactivity, family history of cardiovascular disease and social status (Woodward et al. 2007). This has led to the search for additional diagnostic tools. In an effort to improve coronary heart disease risk prediction, several new biomarkers are effective indicators of high risk of cardiovascular disease namely high-sensitivity C-reactive protein (hs-CRP), serum amyloid A, interleukin-6, homocysteine, soluble intercellular adhesion molecules (sICAM), soluble vascular adhesion molecules (sVCAM), vascular endothelial growth factor (VEGF), thrombomodulin and natriuretic peptides have been evaluated as potential adjuncts to lipid screening in primary prevention. Of these, to date, over a dozen prospective epidemiological studies carried out by individuals with no prior history of cardiovascular disease demonstrate that a single, non-fasting measure of C-reactive protein (CRP) is a strong predictor of future vascular events (Danesh et al. 2000; Hossain et al. 2012; Ridker 2001; Ridker et al. 2000). Most of the clinical studies have also been reported that CRP is an independent risk predictor of atherosclerosis, cardiovascular events, atherothrombosis, hypertension and myocardial infraction (Castell et al. 1990; Devaraj et al. 2003; Hayaishi-Okano et al. 2002; Ross 1999; Verma et al. 2002).

Although hs-CRP and FRS are significantly more effective to predict cardiovascular disease risk, few data are available evaluating which one is better for early risk prediction. To the best of our knowledge, there is no study conducted in the

Bangladeshi population to evaluate the risk prediction of CVD in hyperglycemic subjects. The aim of the present study is to evaluate which one is better in predicting the risk of coronary artery disease (CAD) in hyperglycemic subjects of Bangladesh.

2 Materials and Methods

2.1 Selection of Study Participants

Subjects were selected from OPD (health checkup unit) of Square Hospitals Limited, Dhaka, Bangladesh. A total of 201 study participants with 25–70 years of age are enrolled in this study. Pregnant and lactating mothers and individuals who had a previous and recent history of drug addiction, chronic alcoholism are excluded from the study. Enrolled subjects are non-obese, non-user of anti-inflammatory drugs, anti-hypertensive drugs, statins or estrogen and having no history of renal insufficiency, liver disease, pulmonary disease and rheumatoid arthritis. Information regarding the existing medical conditions and demographic data are collected from the study subjects by a standard questionnaire.

2.2 Ethical Permission

Ethical permission for this study is approved by the Institute of Biological Sciences, University of Rajshahi, Bangladesh and informed consent is obtained from all participants. Confidentialities and rights of the study subjects are strictly maintained as per the guideline of the Institute of Biological Sciences.

2.3 Blood Pressure Measurement

The standard protocol for measuring blood pressure recommended by the World Health Organization is used in this study. After study subjects have rested for 20 min or longer, both systolic and diastolic blood pressures (SBP and DBP) are measured three times with a mercury sphygmomanometer with subjects sitting. SBP and DBP are defined at the first and fifth phases of Korotkoff sounds, respectively. The average of three measurements is used for the analysis.

2.4 Collection of Specimen

Blood samples are collected from the study participants after overnight fasting by venipuncture in two different plastic vacuum tubes—4 ml red top tube contains clot activator (silica) and 2 ml grey top tube contains NaF + K_3EDTA. Samples were allowed to form clot at room temperature and subsequently centrifuged at 1600 × g for 15 min.

2.5 Biochemical Investigations

Plasma glucose, total cholesterol (TC), triglycerides (TG) and HDL-cholesterol (HDL-C) were measured by enzymatic assay kit according to the manufacturer's protocol on a Johnson & Johnson VITROS 350 analyzer (Ortho-Clinical Diagnostics, Rochester, USA). LDL-cholesterol (LDL-C) is calculated by the Friedewald equation. hs-CRP and lipoprotein(a) are measured by particle-enhanced immunonephelometry according to the manufacturer's protocol on a BN ProSpec analyzer (Dade Behring Marburg GmbH, Germany). Apolipoprotein A-1 and apolipoprotein B are simultaneously measured on this device by immunonephelometry. Samples are handled in identical and in blinded fashion throughout the study. CAD risk of study subjects is estimated with the use of Framingham Risk Score (FRS) (Expert Panel on Detection, Evaluation, Treatment of High Blood Cholesterol in Adults 2001; Koenig et al. 2004).

2.6 Statistics

All statistical analyses are performed using software of Statistical Packages for Social Sciences (SPSS) for Windows 15.0 (SPSS Inc., Chicago, Illinois). Study subjects are categorized into three groups based on the fasting plasma glucose (FPG) recommended by American Diabetes Association (ADA) (American Diabetes Association 2006; Committee and on the Diagnosis Classification of Diabetes Mellitus 2003) into normal fasting glucose (NFG) (FPG < 5.6 mmol/L), impaired fasting glucose (IFG) (FPG 5.6 – <7 mmol/L) and diabetes (FPG ≥ 7 mmol/L) group. Descriptive characteristics of the study subjects are performed by several statistical tools. Data are presented as mean ± SD (95% confidence interval), unless otherwise indicated. *p*-values are displayed from one-way ANOVA. Pearson's correlation is used to evaluate the correlation between FPG level and cardiac risk factors. The association between hs-CRP and FPG (continuous variables) or FPG (grouping variables) is performed through linear regression before and after adjusting for covariates. A two-tailed *p*-value of < 0.05 is considered statistically significant.

3 Results

3.1 Descriptive Characteristics of Study Participants

Table 1 shows the descriptive characteristics of the study participants. Of the total 201 study subjects, 91 NFG, 56 IFG and 54 study subjects were diabetes with a mean age range 44.30 ± 8.47, 47.0 ± 9.42 and 52.57 ± 11.94 years, respectively. The average systolic blood pressures (SBP) of the three groups are 114.91, 120.96 and 133.52 mmHg, respectively, that dipper statistically significantly with each other

Table 1 Characteristics of study participants and biochemical parameters according to glycemic status

Variables	Normoglycemia	Hyperglycemia		*p*-value
	NFG	IFG	Diabetes	
n (Male/Female)	91 (75/16)	56 (47/9)	54 (44/10)	
Age (Mean ± SD) year	44.30 ± 8.47	47.0 ± 9.42	52.57 ± 11.94	0.225[a], 0.000[b], 0.023[c]
Systolic blood pressure (mmHg)	114.91 (113.3–116.4)	120.96 (118.9–123.0)	133.52 (131.3–135.6)	0.000[a], 0.000[b], 000[c]
FPG (mmol/L)	5.01 (4.94–5.07)	5.95 (5.86–6.03)	9.44 (8.72–10.16)	0.000[a], 0.000[b], 0.000[c]
Total cholesterol (mg/dL)	178.42 (171.0–185.7)	186.57 (176.2–196.9)	176.19 (163.8–188.5)	0.490[a], 0.985[b], 0.485[c]
Triglycerides (mg/dL)	174.63 (155.0–194.2)	188.68 (168.2–208.5)	194.93 (163.2–226.6)	0.679[a], 0.622[b], 0.982[c]
HDL-cholesterol (mg/dL)	36.95 (35.1–38.8)	35.93 (33.3–38.5)	33.83 (31.8–35.8)	0.891[a], 0.073[b], 0.493[c]
LDL-cholesterol (mg/dL)	106.55 (99.6–113.4)	112.73 (103.7–121.8)	103.13 (94.6–111.6)	0.630[a], 0.898[b], 0.332[c]
Total cholesterol/HDL ratio	5.10 (4.7–5.4)	5.43 (5.0–5.8)	5.36 (4.9–5.7)	0.455[a], 0.682[b], 0.992[c]
hs-CRP (mg/L)	1.78 (1.39–2.17)	3.28 (2.30–4.24)	5.23 (3.88–6.58)	0.017[a], 0.000[b], 0.041[c]
Apolipoprotein A-1 (g/L)	1.21 (1.17–1.25)	1.16 (1.09–1.22)	1.17 (1.11–1.22)	0.460[a], 0.421[b], 1.00[c]
Apolipoprotein B (g/L)	0.93 (0.89–0.96)	0.99 (0.91–1.05)	0.92 (0.86–0.98)	0.365[a], 0.999[b], 0.441[c]
Lipoprotein(a) (g/L)	0.26 (0.19–0.32)	0.23 (0.17–0.27)	0.26 (0.19–0.32)	0.822[a], 1.00[b], 0.812[c]

Source Authors' own data analysis

Values are mean (95% confidence interval), unless otherwise indicated. [a]*p*-value comparison between NFG and IFG groups; [b]*p*-value comparison between NFG and diabetes subjects; [c]*p*-value comparison between IFG and diabetes subjects

($p < 0.001$). TG levels of NFG, IFG and diabetic subjects are higher than the upper normal TG level (<150 mg/dl). There are no significant differences between groups with respect to TC, TG, LDL-C, HDL-C and TC/HDL-C ratio. Apolipoprotein A-1, apolipoprotein B and lipoprotein(a) are not significantly different among the three groups. The average values of hs-CRP of the study subjects in NFG, IFG and diabetes group are found to be 1.78, 3.28 and 5.23 ml/L, respectively, which differ statistically significantly with each other.

3.2 Association of Fasting Plasma Glucose and hs-CPR with Cardiac Risk Factors

Table 2 shows the correlation of cardiac risk parameters that are linked to FRS with the fasting plasma glucose concentration. Among the components of FRS (age, smoking status, SBP, TC and HDL-C), age and SBP are positively and significantly ($r = 0.290$, $p < 0.001$ and $r = 0.615$, $p < 0.001$) while HDL-C is negatively and significantly correlated ($r = -0.137$, $p < 0.05$) with an increase in concentration of FPG. The estimated Framingham 10-year risk of CAD (%) is positively and significantly associated with an increase in FPG ($r = 0.286$, $p < 0.001$). A significant increase in serum hs-CRP levels was observed with an increase in concentrations of FPG ($r = 0.295$, $p < 0.001$). Pearson's correlation analysis is also performed to evaluate the association between the serum concentration of hs-CRP and the estimated Framingham 10-year risk of CAD (%). The result reveals that the estimated Framingham 10-year risk of CAD (%) is positively and significantly associated with an increase in concentration of hs-CRP ($r = 0.210$, $p < 0.001$) (data not shown).

Table 2 Pearson's correlation coefficients between fasting plasma glucose level and cardiac risk factors

Parameters		All subjects	
		Correlation coefficients (r) with FPG	p-value
FRS components	Age	0.290	0.000
	Systolic blood pressure (mmHg)	0.615	0.000
	Smoking status	0.088	0.219
	Total cholesterol (mg/dL)	0.044	0.593
	HDL-cholesterol (mg/dL)	−0.137	0.052
Framingham 10-year risk of CAD (%)		0.286	0.000
hs-CRP (mg/L)		0.295	0.000

Source Authors' own data analysis

3.3 *Serum hs-CRP Level and FPG*

Table 3 shows the association between fasting plasma glucose and high-sensitivity C-reactive protein through linear regression analysis. Before and after adjusting for covariates (age, sex, smoking status, TC and HDL-C), we found that FPG is positively and significantly associated with hs-CRP level. Later, to investigate the concentration–response relationship between FPG concentrations and hs-CRP level, we evaluated the hs-CRP level in the three groups (such as NFG, IFG and diabetes) of the study subjects. Interestingly, we found that hs-CRP levels (Table 4) are significantly increased in the higher FPG groups (IFG and Diabetes) before and after adjustment for covariates (age, sex, smoking status, TC and HDL-C) compared with NFG group (reference group). We categorized serum concentrations of hs-CRP into three groups according to previously defined cut points: low (<1 mg/L), medium (1–3 mg/L) and high (>3 mg/L) to evaluate inflammatory condition of the subjects. We observed that the percentage of higher hs-CRP level increases as increasing fasting plasma glucose (Table 5).

4 Discussion

Inflammation plays an important role in all stages of the atherosclerotic process, from the onset of initial lesions to plaque progression and complications (Ridker et al. 2000). Several commercial assays for inflammatory markers have become available to predict the incident of myocardial infraction, stroke, peripheral arterial disease and sudden cardiac death. In terms of clinical application, CRP seems to be a stronger predictor of cardiovascular events than lipid profile, and it adds prognostic information at all levels of calculated Framingham Risk and at all levels of the metabolic syndrome (Liuzzo et al. 1994; Ridker 2001). Type 2 diabetes mellitus (DM) has been recognized as a strong risk factor for CAD due to chronic oxidative, which, in turn, is the leading cause of mortality and morbidity in diabetic patients compared with their non-diabetic counterpart (Fornengo et al. 2006; Nathan 1993). Although recent studies have shown that higher concentrations of hs-CRP in blood are associated with an increased cardiovascular risk in type 2 DM patients without previous history of cardiovascular disease (Schulze et al. 2004; Soinio et al. 2006). Data on the prognostic value of hs-CRP in type 2 DM with FRS correlation are limited. Silent myocardial ischemia is responsible for more delayed diagnosis of CAD and poorer prognosis than anginal episodes, early detection and routine screening of CAD with simple tool have become important and desirable for diabetic population (Kharlip et al. 2006; Zellweger 2006). The current study indicates that several soluble biomarkers could be predictive for CAD in patients with type 2 diabetes, among which serum levels of hs-CRP appeared to be useful in clinical practice. Interestingly, we found that hs-CRP level is positively correlated with the fasting plasma glucose and this correlation is statistically significant.

Table 3 Association between fasting plasma glucose and high-sensitivity C-reactive protein by regression analysis

Dependent variables	Independent variables	Before adjusting covariates				After adjusting covariates[a]			
		Coefficient (95% CI)	*p*-value (t-test)	R^2	*p*-value† (F-test)	Coefficient (95% CI)	*p*-value (t-test)	R^2	*p*-value† (F-test)
FPG	hs-CRP	0.475 (0.256–0.694)	<0.001	0.085	<0.001	0.399 (0.174–0.624)	<0.001	0.194	<0.001

Source Authors' own data analysis

[a]Adjusted for age, sex, smoking, total cholesterol and HDL-cholesterol; degree of freedom (df) before and after adjustment for covariates were (1, 198) and (1, 193), respectively; †Goodness of-fit of the overall model

Table 4 Dose-response relationship of hs-CRP in the NFG, IFG and diabetes study subjects by regression analysis

Dependent variables	Independent variables	Before adjusting covariates				After adjusting covariates[a]			
		Coefficient (95% CI)	*p*-value (t-test)	R^2	*p*-value† (F-test)	Coefficient (95% CI)	*p*-value (t-test)	R^2	*p*-value† (F-test)
hs-CRP	FPG < 5.6 mmol/L (NFG)	–	–	–	–	–	–	–	–
	FPG 5.6 – <7.0 mmol/L (IFG)	1.82 (0.64–2.99)	<0.01	0.153	<0.001	1.50 (0.35–2.66)	<0.05	0.246	<0.001
	FPG ≥7.0 mmol/L (Diabetes)	3.56 (2.37–4.75)	<0.001			3.23 (1.98–4.47)	Source: Authors' own data analysis		

Source Authors' own data analysis

Fasting plasma glucose (FPG) < 5.6 mmol/L is used as reference group. [a]Adjusted for age, sex, smoking, total cholesterol and HDL-cholesterol; degrees of freedom (df) before and after adjustment for covariates were (2, 197) and (2, 192), respectively; †Goodness-of-fit of the overall model

Table 5 Percentage of hs-CRP cut-of-point in NFG, IFG and diabetes groups

hs-CRP (mg/L)	NFG (%)	IFG (%)	Diabetes (%)
<1 mg/L	48.4	25.0	16.7
1–3 mg/L	37.4	39.3	25.9
>3 mg/L	14.3	35.7	57.4

Source Authors' own data analysis

Our study is in good agreement with previous report has shown that higher concentrations of hs-CRP are associated with an increased cardiovascular risk in type 2 DM patients without previous history of cardiovascular disease (Schulze et al. 2004; Soinio et al. 2006). Researchers did not show any prognostic value of hs-CRP in diabetic population (Ridker et al. 2000; Sabatine et al. 2007). Our data indicate an important inflammatory component of diabetes and its proatherosclerotic role. In fact, inflammation associated with diabetes is likely to play a key role in the early initiation and aggressive progression of atherosclerosis. Measurement of markers of inflammation has been proposed as a method to improve the prediction of the risk of these events. Among the inflammatory markers, hs-CRP is a stronger biomarker for the risk prediction of cardiovascular diseases.

Among the limitations of our study, direct measures of adiposity are not used, in addition to the relatively small number of study subjects enrolled, is not a follow-up study, which does not allow evaluation of the actual prognostic role of hs-CRP in type 2 DM. Among the several limitations, we used several statistical tools to evaluate the association of hs-CRP level with the level of fasting plasma glucose before and after adjusting for covariates. Interestingly, we found that hs-CRP level is significantly increased in the higher FPG (IFG and Diabetes) groups before and after adjustment by covariates compared to NFG (reference group). Another way to estimate the risk of CAD is the calculation of FRS. But this calculation requires individual's information of age, systolic blood pressure, total cholesterol, HDL-cholesterol, smoking status and gender specificity. But in our study, without considering these factors, only hs-CRP measurement may be useful as an independent marker for assessing likelihood of recurrent events of atherosclerosis and coronary artery disease with an increase in fasting plasma glucose.

5 Conclusion

Several commercial assays for inflammatory markers have become available. As a consequence of the expanding research base and availability of assays, the number of inflammatory marker tests ordered by clinicians for CVD risk prediction has grown rapidly. In this study, we observed that, in terms of clinical application, hs-CRP is a stronger predictor of atherosclerotic coronary artery disease in hyperglycemic subjects, and it adds prognostic information at all levels of calculated Framingham

Risk. Thus, the results of this study may potentially be helpful to assess the risk and magnitude of the CAD induced by hyperglycemia.

References

American Diabetes Association (2006) Diagnosis and classification of diabetes mellitus. *Diabetes Care, 29(Suppl 1)*, S43–48.

Anderson, K. M., Odell, P. M., Wilson, P. W., & Kannel, W. B. (1991). Cardiovascular disease risk profiles. *American Heart Journal, 121*, 293–298.

Brindle, P., Beswick, A., Fahey, T., & Ebrahim, S. (2006). Accuracy and impact of risk assessment in the primary prevention of cardiovascular disease: A systematic review. *Heart, 92*, 1752–1759.

Castell, J. V., Gómez-Lechón, M. J., David, M., Fabra, R., Trullenque, R., & Heinrich, P. C. (1990). Acute-phase response of human hepatocytes: Regulation of acute-phase protein synthesis by interleukin-6. *Hepatology, 12*, 1179–1186.

Committee and on the Diagnosis and Classification of Diabetes Mellitus, 2003 The Expert Committee on the Diagnosis and Classification of Diabetes Mellitus (2003) Follow-up report on the diagnosis of diabetes mellitus. *Diabetes Care, 26*, 3160–3167.

Danesh, J., Whincup, P., Walker, M., Lucy, L., Andrew, T., Paul, A., Gallimore, J. R., & Mark, B. P. (2000). Low grade inflammation and coronary heart disease: Prospective study and updated meta-analyses. *British Medical J, 321*, 199–204.

Devaraj, S., Xu, D. Y., & Jialal, I. (2003). C-reactive protein increases plasminogen activator inhibitor-1 expression and activity in human aortic endothelial cells: Implications for the metabolic syndrome and atherothrombosis. *Circulation, 107*, 398–404.

Expert Panel on Detection and Evaluation, and Treatment of High Blood Cholesterol in Adults 2001 Expert Panel on Detection, Evaluation, and Treatment of High Blood Cholesterol in Adults (2001) Executive summary of the third report of the National Cholesterol Education Program (NCEP) expert panel on detection, evaluation and treatment of high blood cholesterol in adults (Adult Treatment Panel III). *JAMA, 285*, 2486-2897.

Fornengo, P., Bosio, A., Epifani, G., Pallisco, O., Mancuso, A., & Pascale, C. (2006). Prevalence of silent myocardial ischaemia in new-onset middle-aged Type 2 diabetic patients without other cardiovascular risk factors. *Diabetic Medicine, 23*, 775–779.

Hayaishi-Okano, R., Yamasaki, Y., Katakami, N., Ohtoshi, K., Goragawa, S. I., Kuroda, A., Matsuhisa, M., Kosugi, K., Nishikawa, N., Kajimoto, Y., & Hori, M. (2002). Elevated C-reactive protein associates with early-stage carotid atherosclerosis in young subjects with type 1 diabetes. *Diabetes Care, 25*, 1432–1438.

Hossain, E., Islam, K., Yeasmin, F., Karim, M. R., Rahman, M., Agarwal, S., Hossain, S., Aziz, A., Mamun, A. A., Sheikh, A., Haque, A., Hossain, M. T., Hossain, M., Haris, P. I., Ikemura, N., Inoue, K., Miyataka, H., Himeno, S., & Hossain, K. (2012). Elevated levels of plasma Big endothelin-1 and its relation to hypertension and skin lesions in individuals exposed to arsenic. *Toxicology and Applied Pharmacology, 259*, 187–194.

Kharlip, J., Naglieri, R., Mitchell, B. D., Ryan, K. A., & Donner, T. W. (2006). Screening for silent coronary heart disease in type 2 diabetes: Clinical application of American Diabetes Association guidelines. *Diabetes Care, 29*, 692–694.

Koenig, W., Löwel, H., Baumert, J., & Meisinger, C. (2004). C-reactive protein modulates risk prediction based on the Framingham score: Implications for future risk assessment: Results from a large cohort study in southern Germany. *Circulation, 109*, 1349–1353.

Laakso, M., & Kuusisto, J. (1996). Epidemiological evidence for the association of hyperglycaemia and atherosclerotic vascular disease in non-insulin-dependent diabetes mellitus. *Annals of Medicine, 28*, 415–418.

Liuzzo, G., Biasucci, L. M., Gallimore, J. R., Grillo, R. L., Rebuzzi, A. G., Pepys, M. B., & Maseri, A. (1994). The prognostic value of C-reactive protein and serum amyloid a protein in severe unstable angina. *New England Journal of Medicine, 331*, 417–424.

Nathan, D. M. (1993). Long-term complications of diabetes mellitus. *New England Journal of Medicine, 328*, 1676–1685.

Ridker, P. M. (2001). High-sensitivity C-reactive protein: Potential adjunct for global risk assessment in the primary prevention of cardiovascular disease. *Circulation, 103*, 1813–1818.

Ridker, P. M., Hennekens, C. H., Buring, J. E., & Rifai, N. (2000). C-reactive protein and other markers of inflammation in the prediction of cardiovascular disease in women. *New England Journal of Medicine, 342*, 836–843.

Ross, R. (1999). Atherosclerosis—an inflammatory disease. *New England Journal of Medicine, 340*, 115–126.

Ruderman, N. B., Gupta, S., & Sussman, I. (1992). Hyperglycemia, diabetes, and vascular disease: An overview. In N. Ruderman, J. Williamson, & M. Brownlee (Eds.), *Hyperglycemia, Diabetes, and Vascular Disease. Clinical Physiology Series* (pp. 3–20). New York: Springer.

Sabatine, M. S., Morrow, D. A., Jablonski, K. A., Rice, M. M., Warnica, J. W., Domanski, M. J., Hsia, J., Gersh, B. J., Rifai, N., Ridker, P. M., Pfeffer, M. A., Braunwald, E., & Investigators, P. E. A. C. E. (2007). Prognostic significance of the Centers for Disease Control/American Heart Association high-sensitivity C-reactive protein cut points for cardiovascular and other outcomes in patients with stable coronary artery disease. *Circulation, 115*, 1528–1536.

Schulze, M. B., Rimm, E. B., Li, T., Rifai, N., Stampfer, M. J., & Hu, F. B. (2004). C-reactive protein and incident cardiovascular events among men with diabetes. *Diabetes Care, 27*, 889–894.

Soinio, M., Marniemi, J., Laakso, M., Lehto, S., & Rönnemaa, T. (2006). High-sensitivity C-reactive protein and coronary heart disease mortality in patients with type 2 diabetes: A 7-year follow-up study. *Diabetes Care, 29*, 329–333.

Steiner, G. (1985). Atherosclerosis, the major complications of diabetes. *Advances in Experimental Medicine and Biology, 189*, 277–297.

Verma, S., Wang, C. H., Li, S. H., Dumont, A. S., Fedak, P. W. M., Badiwala, M. V., Dhillon, B., Weisel, R. D., Li, R. K., Mickle, D. A. G., & Stewart, D. J. (2002). A self-fulfilling prophecy: C-reactive protein attenuates nitric oxide production and inhibits angiogenesis. *Circulation, 106*, 913–919.

Wilson, P. W., D'Agostino, R. B., Levy, D., Belanger, A. M., Silbershatz, H., & Kannel, W. B. (1998). Prediction of coronary heart disease using risk factor categories. *Circulation, 97*, 1837–1847.

Woodward, M., Brindle, P., & Tunstall-Pedoe, H. (2007). Adding social deprivation and family history to cardiovascular risk assessment: The ASSIGN score from the Scottish Heart Health Extended Cohort (SHHEC). *Heart, 93*, 172–176.

Yusuf, S., Hawken, S., Ounpuu, S., Dans, T., Avezum, A., Lanas, F., et al. (2004). Effect of potentially modifiable risk factors associated with myocardial infarction in 52 countries (the Interheart study): Case-control study. *Lancet, 364*, 937–952.

Zellweger, M. J. (2006). Prognostic significance of silent coronary artery disease in type 2 diabetes. *Herz, 31*, 240–245.

Identification of Outliers in Gene Expression Data

Md. Manzur Rahman Farazi and A. H. M. Rahmatullah Imon

Abstract Identification of outliers is a big challenge in big data although it has drawn a great deal of attention in recent years. Among all big data problems, the detection of outliers in gene expression data warrants extra attention because of its inherent complexity. Although a variety of outlier detection methods are available in the literature, Tomlins et al. (Tomlins et al. Science 310:644–648, 2005) argued that traditional analytical methods, for example, a two-sample t-statistic, which search for common activation of genes across a class of cancer samples, will fail to detect cancer genes, which show differential expression in a subset of cancer samples or cancer outliers. They developed the cancer outlier profile analysis (COPA) method to detect cancer genes and outliers. Inspired by the COPA statistic, some authors have proposed other methods for detecting cancer-related genes with cancer outlier profiles in the framework of multiple testing (Tibshirani and Hastie Tibshirani and Hastie Biostatistics 8:2–8, 2007; Wu Wu Biostatistics 8:566–575, 2007; Lian Lian Biostatistics 9:411–418, 2008; Wang and Rekaya Wang and Rekaya Biomarker Insights 5:69–78, 2010). Such cancer outlier analyses are affected by many problems especially if there is an outlier in the dataset then classical measures of location and scale are seriously affected. So the test statistic using these parameters might not be appropriate to detect outliers. In this study, we try to robustify one existing method. We propose a new technique called expressed robust t-statistic (ERT) for the identification of outliers. The usefulness of the proposed methods is then investigated through a Monte Carlo simulation.

This chapter draws some material from MS Dissertation of the first author and the work was supervised by the second author. The dissertation was duly published by the Graduate School of Ball State University, USA in 2015. The authors have received necessary permission from the publisher.

Md. M. R. Farazi
Medical College of Wisconsin, Shorewood, WI 53211, USA
e-mail: mfarazi@mcw.edu

A. H. M. Rahmatullah Imon (✉)
Department of Mathematical Sciences, Ball State University, Muncie, IN 47306, USA
e-mail: rimon@bsu.edu

B. K. Sinha et al. (eds.), *Data Science and SDGs*,
https://doi.org/10.1007/978-981-16-1919-9_11

Keywords Gene expression · Cancer outlier profile · Multiple outlier · Masking · Swamping · Robust statistics

1 Introduction

Statistical data may comprise some unusually small or large observations, so-called outliers. Measurement and execution errors are the main sources of outliers. But that is not all. Outliers may occur due to the inherent variability of the data, i.e., the natural feature of the population may be uncontrollable. Determining whether a dataset contains one or more outliers is a challenge commonly faced in applied statistics. This is a mostly difficult mission if the properties of the underlying population are not known. However, in many empirical investigations, the assumption that the data come from a particular population is too restrictive or unrealistic. Although outliers are often considered as an error or noise, they may convey important information. Detected outliers are candidates for peculiar data that may otherwise adversely lead to model misspecification, biased parameter estimation and incorrect results. It is therefore important to identify them prior to modeling and analysis. Outlier detection methods have been suggested for numerous applications that include clinical trials. An excellent review of different aspects of outlier detection is available in Hawkins (1980), Barnett and Lewis (1994), and Hadi et al. (2009). When cancer data can be modeled by a logistic regression or by any other generalized linear models, outliers can be identified using generalized standardized Pearson residuals (see Imon and Hadi 2009), the generalized difference in fits (see Nurunnabi et al. 2010) and distance from the median (see Imon and Hadi 2013).

In statistical data, the concept of outliers is global. Outliers are the observations that fail to match with a global pattern (parent distribution). But most of the time big data do not obey any common global pattern and observations may come in clusters where the concept of majority and minority simply do not work here as shown in Fig. 1. Here it is not easy to tell which portion of the data is major.

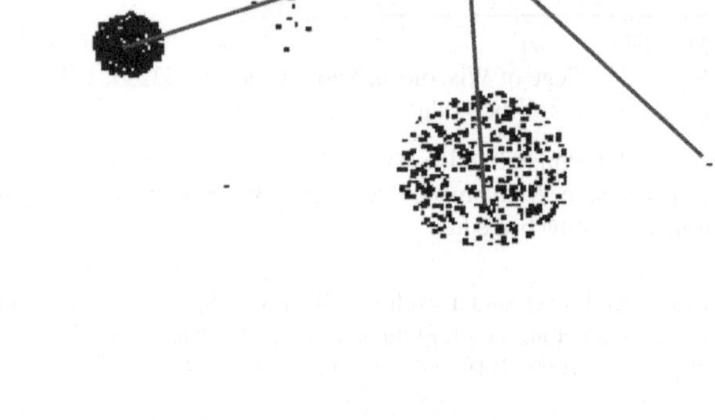

Fig. 1 Outliers in data clusters. *Source* Imon and Hadi (2020)

However, there exist four major clusters here, and in addition to those four clusters, there exist few scatter data points, which fail to accommodate with any of the four clusters. Cluster-based outlier detection methods are widely used nowadays (see Aggarwal and Yu 2000). Among the cluster-based outliers, the DB(ε,π)-outliers and grid-based outliers, index-based outliers, nested loop-based outliers proposed by Knorr and Ng (1997, 1998), the local outlier factor (LOF) proposed by Breunig et al. (1999), k-nearest neighborhood approach proposed by Ramaswamy et al. (2000), resolution-based outlier factor (ROF) proposed by Fan et al. (2006) and spatial robust z proposed by Hadi and Imon (2018) have become popular. Aggarwal and Yu (2000) pointed out that the distance between any pair of data points in high-dimensional space is so similar that either each data point or none data point can be viewed as an outlier if the concepts of proximity are used to define outliers. As a result, using traditional Euclidean distance function cannot effectively get outliers in high-dimensional dataset due to the averaging behavior of the noisy and irrelevant dimensions. In addition to the methods described above, there are several tree-based unsupervised learning algorithms like 'isolation forest' proposed by Liu et al. (2008) to identify both univariate and multivariate outliers. The issue of robustness of spatial outlier methods in the presence of multiple outliers is discussed by Filzmoser (2014). These two approaches are based on Mahalanobis or robust distances computed in each neighborhood using a common estimation of the covariance matrix.

Since a good number of classification and/or outlier detection methods are available in the literature the question might arise, do we still need a new one? The answer to this question is, perhaps, yes. The simple classification methods suffer from masking and swamping. On the other hand, sophisticated outlier detection methods are computationally very intensive and hence are not very friendly for big data. For this reason, we need a classification technique that is easy to execute and robust at the same time. Tomlins et al. (2005) led the way by developing an outlier detection method designed exclusively for the gene expression data. In the next section, we will see that this method and all the subsequent methods developed in this area contain very simple measures such as median and quartiles. The rapid developments of technologies that generate arrays of gene data enable a global view of the transcription levels of hundreds of thousands of genes simultaneously. The outlier detection problem for gene data has its importance but together with the difficulty of high dimensionality. The scarcity of data in high-dimensional space makes each point a relatively good outlier in the view of traditional distance-based definitions. Thus, finding outliers in high-dimensional data is more complex.

Microarray technology is used in a wide variety of settings for detecting differential gene expression. Classic statistical issues such as appropriate test statistics, sample size, replicate structure, statistical significance and outlier detection enter into the design and analysis of gene expression studies. Adding to the complexity is the fact that the number of samples I in a microarray experiment is inevitably much less than the number of genes J under investigation and that J is often on the scale of tens of thousands, thus creating a tremendous multiple testing burden. Fundamental to the task of analyzing gene expression data is the need to identify genes whose patterns of expression differ according to phenotype or experimental condition. Gene expression is a well-coordinated system, and hence measurements on different genes are in general not independent. Given more complete knowledge of the specific interactions and transcriptional controls, it is conceivable that meaningful comparisons between samples can be made by considering the joint distribution of specific sets of genes. However, the high dimension of gene expression space prohibits a comprehensive exploration, while the fact that our understanding of biological systems is only in its infancy means that, in many cases, we do not know which relationships are important and should be studied. In current practice, differential expression analysis will, therefore, at least start with a gene-by-gene approach, ignoring the dependencies between genes. A simple approach is to select genes using a fold-change criterion. This may be the only possibility in cases where no, or very few replicates, are available. An analysis solely based on fold change, however, does not allow the assessment of the significance of expression differences in the presence of biological and experimental variation, which may differ from gene to gene. This is the main reason for using statistical tests to assess differential expression. Generally, one might look at various properties of the distributions of a gene's expression levels under different conditions, though most often location parameters of these distributions such as the mean are considered. Parametric test, such as the t-test, is commonly used. Parametric tests usually have a higher power if the underlying model assumptions, such as normality in the case of the t-test, are at least approximately fulfilled. The presence of outliers may often destroy normality patterns so it is essential to identify outliers in gene expression data before any further statistical analysis.

2 Existing Outlier Detection Methods for the Gene Expression Data

It is well known that genetic translocations occur in cancer cells, and that these translocations can result in the up-regulation of oncogenes that may affect the progression of cancer. This translocation can happen between the activating gene and multiple oncogenes. Since a given translocation is only likely to occur once per sample, if there were multiple partners for a given activating gene, we would expect to see certain cancer samples with a high expression of say, gene *A*, whereas other cancer samples might have high expression of gene *B*, but these samples would be mutually exclusive. In addition, we would expect that the normal samples would not have high expression for neither gene *A* nor *B*. We can use this idea to both pre-filter genes as well as finding interesting genes that may be involved in translocations.

Assuming case–control microarray data were generated for detecting differentially expressed genes consisting of n samples from a normal group and m samples from a cancer group. Let be the expression value for gene $i = (1, 2, \ldots, p)$ and sample $j = (1, 2, \ldots, n)$ in the normal group and be the expression value for gene $\mathrm{i} = (1, 2, \ldots, \mathrm{p})$ and sample $j = (1, 2, \ldots, m)$ in the cancer group. In this study, and without loss of generality, we are only interested in one-sided tests where the activated genes from cancer samples are overexpressed or upregulated.

t-statistic

The most widely used method for detecting differential gene expression in comparative microarray studies is the two-sample *t*-statistic. A gene is determined to be significant if the absolute *t*-value exceeds a certain threshold *c*, which is usually determined by its corresponding *p*-value or false discovery rate. The two-condition *t*-statistic for gene *i* is defined by

$$t_i = \frac{\bar{y}_i - \bar{x}_i}{s_i} \tag{1}$$

where $\bar{y}_i$ is the mean expression value in cancer samples, $\bar{x}_i$ is the mean expression value in normal samples for gene *i* and s_i is the pooled standard error estimate given by

$$s_i = \sqrt{\frac{\sum_{i=1}^{n}(x_i - \bar{x}_i)^2 + \sum_{i=1}^{m}(y_i - \bar{y}_i)^2}{n+m-2}}. \tag{2}$$

The *t*-statistic is powerful when most cancer samples are activated.

Cancer Outliers Profile Analysis (COPA) Statistic

Tomlins et al. (2005) introduced the cancer outlier profile analysis (COPA) method for detecting cancer genes, which are differentially expressed in a subset of disease samples

$$COPA_i = \frac{q_r(y_{ij}) - Med_i}{MAD_i} \tag{3}$$

where $q_r(.)$ is the *r*th percentile of the expression data, and medi is the median expression value for all samples. *MADi* is the median absolute deviation for the complete sample. For the complete sample u_i, $i = 1, 2, \ldots, m+n$,

$$MAD_i = 1.4826Med\{|u_i - Med(u_i)|\}. \tag{4}$$

Heterogeneous patterns of oncogene activation were observed in the majority of cancer types considered in their studies. Thereafter, several further studies in this direction have been proposed.

Outliers Sums (OS) Statistic

COPA statistic uses a fixed *r*th sample percentile, which is determined by users. This limitation was overcome by the outlier sums (OS) statistic defined by Tibshirani and Hastie (2007) as

$$OS_i = \frac{\sum_{y_{ij} \in R_i} (y_{ij} - Med_i)}{MAD_i} \tag{5}$$

where

$$R_i = \{y_{ij} : y_{ij} > Q_3 + IQR\}. \tag{6}$$

Here $IQR = Q3 - Q1$, $Q1$ is the first quartile and $Q3$ is the third quartile of the combined sample.

Outliers Robust t (ORT) Statistic

Wu (2007) modified the OS statistic by changing the definition of R_i. His proposed outlier's robust *t* (ORT) statistic is defined as

$$ORT_i = \frac{\sum_{y_{ij} \in R_i^*} (y_{ij} - Med(x_i))}{MAD_i^*} \tag{7}$$

where

$$R_i^* = \{y_{ij} : y_{ij} > Q_3(x) + IQR(x)\} \tag{8}$$

and

$$MAD_i^* = 1.4826 Med\{|x_i - Med(x_i)|, |y_i - Med(y_i)|\}. \tag{9}$$

It is clear from the above expressions that (4) and (6) are computed on the combined sample of *x* and *y*, (8) is computed on sample *x* and (9) is computed on *x* and *y* separately. COPA and OS statistics were derived from the *t*-statistic by replacing the mean and standard errors used in the *t*-statistic with the median and median absolute deviations, respectively. ORT has been proposed as a more robust statistic that utilizes the absolute difference of each expression value from the median instead of the squared difference of each expression value from the average.

Maximum Ordered Subset t (MOST) Statistic

Lian (2008) argued that OS and ORT statistics used arbitrary outliers and proposed the maximum ordered subset *t* (MOST) statistic, which consider all possible values for outlier thresholds. The MOST procedure requires cancer sample expression data to be sorted in descending order and we calculate the following statistic

$$MOST_i = \max_{1 \le k \le m} \left(\frac{\sum_{1 \le j \le k} \frac{(y_{ij} - Med(x_i))}{MAD_i^*} - \mu_k}{\delta_k} \right). \tag{10}$$

Here μ_k and δ_k are obtained from the order statistics of m samples generated from a standard normal distribution and are used to make different values of the statistic comparable for different values of k.

The Least Sum of Ordered Subset Variance t-Statistic (LSOSS)

Wang and Rekaya (2010) proposed a new method named least sum of ordered subset square t-statistic (LSOSS). In LSOSS, mean expression values in normal and cancer samples were considered instead of median expression values. Because if outliers are present among cancer samples, the distribution of gene expression values in cancer samples will have two peaks. The higher peak corresponds to activated samples while the lower peak indicates inactivated samples. Consequently, this outlier issue can be addressed through the idea of detecting a "change point" or "break point" in the ordered gene expression values of the cancer group. For each gene i, the expression levels in cancer samples are sorted in descending order and then divided into two subsets

$$S_{ik1} = \{y_{ij} : 1 \leq j \leq k\}, \quad S_{ik2} = \{y_{ij} : k+1 \leq j \leq m\}$$

For the two subsets, the mean and sum of squares for each gene i are calculated. Let us denote them by

$$\bar{y}_{S_{ik1}} = mean(\{y_{ij} : 1 \leq j \leq k\}), \bar{y}_{S_{ik2}} = mean(\{y_{ij} : k+1 \leq j \leq m\})$$

$$SS_{ik1} = \sum_{1 \leq j \leq k} \left(y_{ij} - \bar{y}_{S_{ik1}}\right)^2,$$

$$SS_{ik2} = \sum_{k+1 \leq j \leq m} \left(y_{ij} - \bar{y}_{S_{ik2}}\right)^2$$

The only issue left to be solved is the value k that divided the two subsets. For that purpose, an exhaustive search was implemented for all possible values ranging from 1 to $m - 1$. The optimum value of k is obtained by minimizing the pooled sum of squares for cancer samples given by

$$\arg \min_{1 \leq k \leq m-1} (SS_{ik1} + SS_{ik2}).$$

Finally, the LSOSS statistic for declaring a gene i with outlier differential expression in case samples is computed as

$$LSOSS_i = k \frac{\bar{y}_{S_{ik1}} - \bar{x}_i}{s_i} \tag{11}$$

where k could be interpreted as the number of outlier samples for gene i.

3 Proposed Outlier Detection Method

We often observe that identification methods fail to identify potential outliers or the methods identify cases as outliers, which are actually not. In masking (false negative), it is said that one outlier masks a second outlier, if the second outlier can be considered as an outlier only by itself, but not in the presence of the first outlier. Thus, after the deletion of the first outlier, the second instance has emerged as an outlier. Masking occurs when a cluster of outlying observations skews the mean and the covariance estimates toward it, and the resulting distance of the outlying point from the mean is small. The opposite effect of masking is called swamping (false positive). In describing the swamping effect, it is said that one outlier swamps a second observation, if the latter can be considered as an outlier only under the presence of the first one. In other words, after the deletion of the first outlier, the second observation becomes a non-outlying observation. Swamping occurs when a group of outlying instances skews the mean and the covariance estimates toward it and away from other non-outlying instances, and the resulting distance from these instances to the mean is large, making them look like outliers.

We observe that most of the outlier detection techniques defined in the previous section contain some non-robust components such as the mean and standard deviations and consequently they may become ineffective in doing their jobs. In our study, we propose three new outlier techniques modifying some of the existing ones. The proposed techniques are described below.

We have seen in (1) that the two-condition t-statistic for gene i is defined by

$$t_i = \frac{\bar{y}_I - \bar{x}_I}{s_i}.$$

Since both $\bar{y}_i$, $\bar{x}_i$, and s_i are non-robust, we propose the expressed robust t (ERT) statistic as

$$ERT_i = \frac{Med_i(y) - Med_i(x)}{MAD_i^*}. \tag{12}$$

Here MAD^*_i is the median absolute deviation as defined in (9). Under the null hypothesis of normality Med(x) $\approx$ Mean(x), Med(y) $\approx$ Mean(y), and $MAD^*_i \approx s_i$ and hence ERT_i follows a Student's t distribution with $m + n - 2$ degrees of freedom.

4 Monte Carlo Comparison of Different Outlier Detection Methods

In this section, we report a Monte Carlo simulation experiment, which is designed to compare the performances of different outlier detection methods in gene expression data. We prefer simulation study because here we definitely know which observations are genuine outliers. It is very cumbersome in case of real data because due to masking and swamping of multiple outliers, it is almost impossible to know which observations are outliers. This simulation study is conducted to compare the performance of the newly proposed expressed robust t (ERT) method with the conventional t-statistic, COPA, OS, ORT, MOST and LSOSS. The simulation was conducted in different situations. To test and check the consistency of the test statistic, we generate gene expression for two groups of samples with different sizes and a variety of situations. In all simulations, we generated $g = 40$ genes. Out of 40 genes, we generated 20 genes considering no differences between normal and tumor group. We generated these 20 genes with uniform conditions for both groups. Further, we generated another 20 genes with two different situations. To distinguish the two groups, for normal sample and tumor sample, we used different ranges. We assume outliers do exist in later 20 genes. The process is done five times by changing the number of normal and tumor sample sizes. For the first set of simulation, we generated $n = 75$ and $m = 25$ as number of samples from normal and tumor groups respectively. For other simulations, we chose ($n = 60$, $m = 40$), ($n = 55$, $m = 45$), ($n = 80$, $m = 20$) and ($n = 90$, $m = 10$). We applied all the existing methods and the proposed method to these simulated data. The results of the number of genes detected as outliers in this simulation experiment are given in Table 1.

Results presented in Table 1 show that all the existing methods including COPA, OS and ORT are not very successful in the identification of outlying genes. Their performance deteriorates when the proportion of samples from cancer group decreases. One possible reason for this could be the fact that most of the commonly used measures include quartiles as a part of the detection tool. But quartiles may

Table 1 Outliers in different methods in simulation study

	$n = 75$ $m = 25$ $g = 40$	$n = 60$ $m = 40$ $g = 40$	$n = 55$ $m = 45$ $g = 40$	$n = 80$ $m = 20$ $g = 40$	$n = 90$ $m = 10$ $g = 40$
T	0	0	0	0	0
COPA	6	5	4	4	2
OS	17	15	9	6	4
ORT	7	5	5	4	4
LSSOS	3	0	0	0	0
ERT	19	20	20	20	20

Source Created by authors

break down if the data contain 25% or more outliers. The performance of t and LSSOS are the worst. Most of the time they fail to identify even a single outlier. The newly proposed ERT performs the best. The methods give consistent results over different simulations.

5 Conclusions

In our study, we propose a new outlier detection technique, ERT, for finding outliers in gene expression data. For each gene, ERT distinguishes the expression values of normal and tumor samples. If any gene is expressed heterogeneously in cancer samples, the mean and variance of gene expression values in cancer samples are overemphasized by the classical t-statistic while ERT uses the robust statistic median and MAD as substitutes of them. The simulation results suggest that the performance of ERT is much better than the existing methods in the identification of outliers in gene expression data.

Acknowledgements The authors express their thanks to the anonymous reviewer for giving some useful suggestions that led to considerable improvement of this manuscript. The authors would like to thank the Ball State University for giving permission to use materials of an MS Thesis written by the first author under the supervision of the second author. The authors also express their thanks to the Editor-in-Chief for his valuable comments and suggestions.

References

Barnett, V., & Lewis, T. B. (1994). *Outliers in statistical data* (3rd ed.). Wiley.

Breunig, M. M., Kriegel, H. P., Ng, & Sander, J. R. (1999). OPTICS-OF: Identifying local outliers. In *Proceedings of the Third European Conference on Principles of Data Mining and Knowledge Discovery* (pp. 262–270).

Fan, H., Zaïane, O. R., Foss, A., & Wu, J. (2006). A nonparametric outlier detection for efficiently discovering top-n outliers from engineering data. In *Proceedings of Pacific-Asia Conference on Knowledge Discovery and Data Mining, Singapore* (pp. 557–566).

Filzmoser, P., Ruiz-, A., & Thomas-, C. (2014). Identification of local multivariate outliers. *Statistical Papers, 55*, 29–47.

Hadi, A. S., & Imon, A. H. M. R. (2018). Identification of multiple outliers in spatial data. *International Journal of Statistical Sciences, 16*, 87–96.

Hadi, A. S., Imon, A. H. M. R., & Werner, M. (2009). Detection of outliers, wiley interdisciplinary reviews. *Computational Statistics, 1*, 57–70.

Hawkins, D. M. (1980). *Identification of outliers*. Chapman and Hall.

Imon, A. H. M. R., & Hadi, A. S. (2013). Identification of multiple high leverage points in logistic regression. *Journal of Applied Statistics, 40*, 2601–2616.

Imon, A. H. M. R., & Hadi, A. S. (2020). Identification of multiple unusual observations in spatial regression. *Journal of Statistics and Applications ((A Special Issue in Honour of Prof. Bimal K Sinha and Prof. Bikas K Sinha).), 18*, 155–162.

Knorr, E., & Ng, R. (1997). A unified notion of outliers: properties and computation. In *Proceedings of the International Conference on Knowledge Discovery and Data Mining* (pp. 219–222).
Knorr, E., and Ng, R. (1998). Algorithms for mining distance-based outliers in large datasets. In *Proceedings of 24th International Conference on Very Large Data Bases* (pp. 392–403).
Lian, H. (2008). MOST: Detecting cancer differential gene expression. *Biostatistics, 9*, 411–418.
Liu, F. T., Ting, K. M., & Zhou, Z. (2008). Isolation forest. In *Eighth IEEE International Conference on Data Mining* (pp. 413–22).
Nurunnabi, A. A. M., Imon, A. H. M. R., & Nasser, M. (2010). Identification of multiple influential observations in logistic regression. *Journal of Applied Statistics, 37*, 1605–1624.
Ramaswamy, S., Rastogi, R., & Shim, K. (2000). Efficient algorithms for mining outliers from large data sets. In *Proceedings of the ACM SIGMOD International Conference on Management of Data* (pp. 427–438).
Tibshirani, R., & Hastie, T. (2007). Outlier sums for differential gene expression analysis. *Biostatistics, 8*, 2–8.
Tomlins, S. A., Rhodes, D. R., & Perner, S. (2005). Recurrent fusion of TMPRSS2 and ETS transcription factor genes in prostate cancer. *Science, 310*, 644–648.
Wang, Y., & Rekaya, R. (2010). LSOSS: Detection of cancer outlier differential gene expression. *Biomarker Insights, 5*, 69–78.
Wu, B. (2007). Cancer outlier differential gene expression detection. *Biostatistics, 8*, 566–575.

Selecting Covariance Structure to Analyze Longitudinal Data: A Study to Model the Body Mass Index of Primary School-Going Children in Bangladesh

Mohammad Ohid Ullah and Mst. Farzana Akter

Abstract In the longitudinal study, the data are collected from the same subject over time and hence the data are correlated. To analyze such data selecting an efficient covariance structure is very important to get better results. Therefore, this article is aimed to select an efficient covariance structure to model the body mass index (BMI) of primary school-going children in Bangladesh. In this study, at first, we have conducted a longitudinal survey to build a cohort of 100 primary school-going children in Sylhet city, Bangladesh. We collected the information from the same children at the initial time (T_0), after 6 months (T_6), after 12 months (T_{12}) and after 18 months (T_{18}). Linear mixed model (LMM) is applied for selecting an efficient covariance structure and then to model the body mass index. To find out a better covariance structure, we used diagonal, unstructured (UN), auto Regressive order 1 (AR1) and compound symmetry (CS) covariance structures in collecting longitudinal data. Observing all the criteria, it is found that the covariance structure compound symmetry ('CS') gives better results for LMM. Finally using the CS covariance structure, initially, we observed that the BMI of male students' is comparatively smaller than female students' (Estimate = -−.04, P-value = 0.03). But overtime, a reverse result is observed at T12 and T18. Taken together, we may conclude that compound symmetry (CS) gives better output to model the body mass index of primary school-going children. As female students are getting more obese, in addition, today's female children are the mothers of the future. Therefore, parents should give concentration to female children to reduce their body weight. This study may be useful for researchers in public health sectors to select a proper covariance structure to analyze their longitudinal data.

Keywords Compound symmetry · Obesity · Longitudinal study · Linear mixed model · Public health

---

M. Ohid Ullah (✉) · Mst. Farzana Akter
Department of Statistics, Shahjalal University of Science and Technology, Sylhet, Bangladesh
e-mail: ohid-sta@sust.edu

B. K. Sinha et al. (eds.), *Data Science and SDGs*,
https://doi.org/10.1007/978-981-16-1919-9_12

1 Introduction

There are many ways to measure obesity, such as: by measuring the level of adiposity, by calculating several indices like weight-for-height index, body mass index (BMI), waist-hip ratio (WHR) and by estimating body fat percentage from skinfold thickness (ST), etc. (Ghesmaty Sangachin et al. 2018). Among these, the widely used measure is the body mass index (World Health Organization 1995). The prevalence of obesity is very high in economically developed countries and many of them have declared it as an epidemic (Hill and Peters 1998; James et al. 2001). However, its prevalence in developing countries cannot be ignored. Furthermore, according to the WHO, the increasing rate of obesity is often faster in developing countries than in the developed countries. In Bangladesh, of the children aged 5–18 years, 10% are overweight while 4% are obese (ICDDR 2013). The findings are alarming considering the size of our young population. A previous study reveals that the percentage of overweight is 1.1% among the children aged 0–5 years in Bangladesh (Onis and Blossner 1998). A cross-sectional study has been previously conducted in Sylhet, Bangladesh, to identify the associated factors of Obesity (Ohid Ullah et al. 2014). To the best of our knowledge, no longitudinal study has ever been conducted to model the body mass index (BMI) of primary school-going children by selecting a proper covariance structure. Covariance structure shows the variance and covariance among the repeated data over different time points.

The data collected from the same individuals over time under different conditions are called longitudinal data. As the longitudinal data are correlated, it is very essential for a comprehensive account of biometrical handling of longitudinal growth data (Tanner 1951). Therefore, independent sample t-test, analysis of variance or regression analysis cannot be applicable for longitudinal data. Correlated data of two time-points (i.e., paired data) can be handled by paired t-test. However, if someone wants to adjust any covariate with paired data then paired t-test is not enough. Therefore, for more than two time-points correlated data (i.e., longitudinal data), linear mixed model is applicable (Keselman et al. 1998; Milliken and Johnson 2004; Molenberghs 2000). To analyze longitudinal data by mixed models, covariance structure is very essential. Unfortunately, still it is not clear how to select covariance structure for linear mixed model. Many researchers have shown several approaches for instance: selection by meaning, by ICs (AIC, BIC, AICC, BICC) and by graphically (Kincaid xxxx). There is still no definite method available to select a covariance structure. Therefore, we aimed to propose an approach to select covariance structure for linear mixed model for modeling BMI of primary school-going children in Bangladesh.

2 Materials and Methods

In this longitudinal study, initially, we randomly have selected a total 100 primary school-going children (aged: 6–10 years) in Sylhet city, Bangladesh to create a cohort. We have measured and recorded their heights and weights using standard measurement protocol. Afterward, we followed them total 18 months. The students were willing to give their information and they were well informed about the purpose of the study. We collected the information from the same children (male and female) at the initial time (T_0), after 6 months (T_6), after 12 months (T_{12}) and after 18 months (T_{18}). The design of the study is as follows:

In Table 1, **Y** indicates the set of variables height and weight for each student. We calculated BMI (body mass index) using height and weight. The formula is given below:

$$\text{BMI} = \text{Weight (kg)}/\ (\text{Height (m)})^2.$$

We used covariance structures: diagonal, unstructured, autoregressive (1) and compound symmetry in this study. Selecting a better covariance structure, linear mixed model (LMM) is applied to model the body mass index by using IBM SPSS program. To select covariance, structure the proposed approach is as follows:

1. Initially plotting correlation matrix of BMI among four time points.
2. Using the appropriate fixed effects, run the linear mixed model considering different covariance structures and collect the values of ICs and compare them (smaller is the better).
3. Collect the estimates and standard errors (SE) of the parameters for different covariance structures and compare them as well. Smaller SE is better.
4. Observing the graph, ICs and SE of the parameters of the model, select an appropriate covariance structure.

Table 1 Design of the study

Gender	Student ID	T_0	T_6	T_{12}	T_{18}
Male	1	Y_{11}	Y_{12}	Y_{13}	Y_{14}
	2	Y_{21}	Y_{22}	Y_{23}	Y_{24}
	⋮	⋮	⋮	⋮	⋮
	m	Y_{m1}	Y_{m2}	Y_{m3}	Y_{m4}
Female	1	Y_{11}	Y_{12}	Y_{13}	Y_{14}
	2	Y_{21}	Y_{22}	Y_{23}	Y_{24}
	⋮	⋮	⋮	⋮	⋮
	n	Y_{n1}	Y_{n2}	Y_{n3}	Y_{n4}

Source Created by the authors

3 Results and Discussion

To analyze the data of this study, at first, we calculated BMI and then plotted a mean profile of BMI over time for males and females. We found that initially, average BMI of females is smaller than males. However, after 12 months and 18 months, BMI of females is observed to be larger than males.

To select a covariance structure, **at the first step**, we have plotted correlation matrix among initial time (T_0) after 6 months (T_6), after 12 months (T_{12}) and after 18 months (T_{18}) is shown in Fig. 1. The figure suggests that every pairwise correlation is of a similar pattern. So CS is a better covariance structure for these data.

At the second step, we collected values of the information criteria (ICs) considering Gender, Time and Time*Gender as fixed factors in the linear mixed model. The IC values suggest that the UN covariance structure gives the smallest values followed by CS. **At the third step**, we have estimated model parameters, the estimate and standard errors (SE) of the estimates. The results indicate that the SE of the interaction term of Gender and Time in CS covariance structure is comparatively smaller than the others and the interaction term is also significant in CS (Table 2). This suggests that CS may be better than others structures.

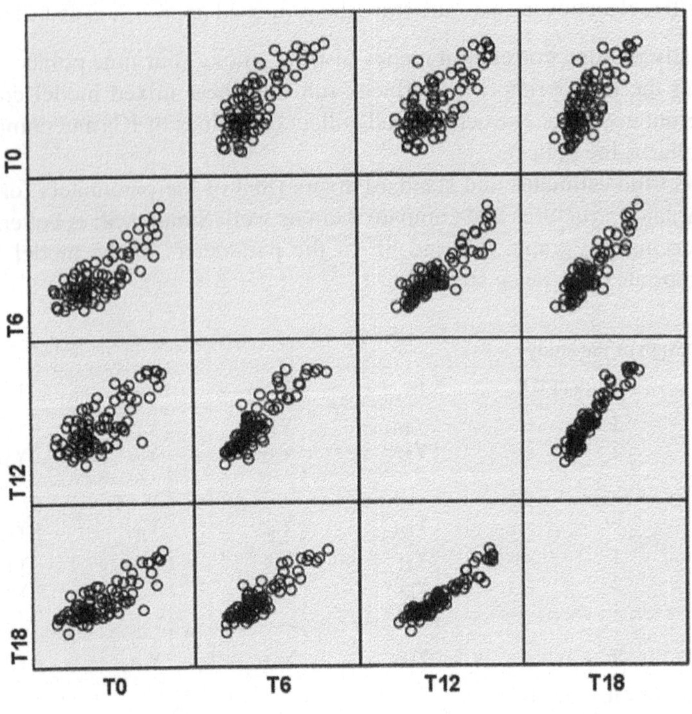

Fig. 1 Correlation matrix of BMI at different time points. *Source*: Created by the authors

Table 2 Estimates and SEs of the parameters of a linear mixed model of the different covariance structures including four time points' data

	DIAG		UN		AR(1)		CS	
Parameter	Estimate	SE	Estimate	SE	Estimate	SE	Estimate	SE
Intercept	0.01	0.34	0.01	0.37	0.01	0.42	0.01	0.42
Gender	0.35	0.53	0.34	0.58	0.40	0.67	0.34	0.66
Time	0.05	0.03	0.04	0.01	0.04	0.02	0.05	0.01
Gender Time	−0.04	0.05	−0.01	0.02	−0.04	0.03	−0.04	0.02

Female is the reference group. *Source* Authors' own calculation by using SPSS.

The above results reveal that CS might be a suitable covariance structure to run linear mixed model for these data. To validate these results, we also have collected ICs and estimate with SE for these four covariance structures in the data of three time points (results not shown here). The results of ICs dictate that UN is better. However, the correlation matrix shows the same pattern up to three time points. So CS may be better like four time points data. The estimation and SEs of the parameters of the three time points data also indicate that for CS structure, SE of the interaction term is smaller than others as well as significant. Including all, the three time points' data also suggested that CS is better. So using CS covariance structure, the final linear mixed model of BMI of primary school-going children is:

$$\mathbf{BMI = 0.01 + 0.34 * Gender + 0.05 * Time - 0.04 * Gender * Time.}$$

We have observed that there is no significant difference between average BMI of males and females. The time has a significant influence on BMI, that is BMI of children is different at different time points as they are growing in age and their BMIs are changed of course. To see the evolution, the most important information in the mixed model is the interaction of gender with time, that is, mixed model is helpful to see the evolution of BMI of males and females over time. The final selected model indicates that BMI of males is smaller (-0.04, $p = 0.03$) compared with females over time. It suggests that the BMI of females is increasing more than males.

As today's children are the parents of the future and mother's health is very important for their child health, so we have to give more concentration on female children in our family. According to WHO, the prevalence of obesity is increasing globally with 300 million people are clinically obese of 1 billion overweight adults. Childhood obesity is already epidemic in some developed countries and rising in developing countries (Obesity and Overweight 2003). In 2013, 42 million infants and young children were overweight or obese, worldwide, and it will turn out to be 70 million by 2025 (World Health Organization 2015). Although the extent of change of BMI has not been found so much in this study, in the future, it may be a major problem. Both parents and existing family members should be more conscious about the horrific effect of obesity and overweight.

4 Conclusion

This study simply shows the evolution of BMI between male and female children over time by selecting a suitable covariance structure to analyze longitudinal data. As there is no definitive approach to select covariance structure for the linear mixed model, we tried to propose an approach to select covariance structure to build a model-observing SEs of the parameters with ICs. We have considered only four covariance structures in this study. More covariance structures in different kinds of study designs as further research are needed for selecting a suitable covariance structure to analyze a linear mixed model. Taken together, we may conclude that to analyze a linear mixed model and to select a covariance structure, it is always better to go a simple way. As we observed that the average BMI of females is increasing more than males, so parents should be more conscious about the health of their female children. This research work may help (i) the policymakers to take proper steps for taking a balanced diet to improve child health in Bangladesh and (ii) it may be useful for researchers in public health sectors to select a proper covariance structure for the linear mixed model to analyze their longitudinal data.

Acknowledgements We would like to thank the SUST Research Center, Sylhet, Bangladesh for providing partial funds for this study.

References

de Onis, M., & Blossner, M. (1998). Prevalence and trends of overweight among preschool children in developing countries. *American Journal of Clinical Nutrition, 72,* 1032–1039.

Ghesmaty Sangachin, M., Cavuoto, A. L., & Wang, Y. (2018). Use of various obesity measurement and classification methods in occupational safety and health research: A systematic review of the literature. *BMC Obesity, 5,* 28.

Hill, J. O., & Peters, J. C. (1998). Environmental contributions to the obesity epidemic. *Science, 280*(5368), 1371–1374.

ICDDR, B. (2013). New ICDDR, B study reveals that 14 out of every 100 children living in an urban area in Bangladesh are overweight. https://www.icddrb.org/dmdocuments/Press%20Release_National%20Obseity%20study_3%20July%202013_NA_edit%20anaheed_revised%20Dec_13.pdf

James, P. T., Leach, R., Kalamara, E., & Shayeghi, M. (2001). The worldwide obesity epidemic. *Obesity Research, 9*(S11), 228S-233S.

Keselman, H. J., Algina, J., Kowalchuk, R. K., & Wolfinger, R. D. (1998). A comparison of two approaches for selecting covariance structures in the analysis of repeated measurements. *Communications in Statistics—Simulation and Compution, 27,* 591–604. https://doi.org/10.1080/03610919808813497

Kincaid C. (2005). *Guidelines for selecting the covariance structure in mixed model analysis.* Portage, MI: COMSYS Information Technology Services, Inc. https://support.sas.com/resources/papers/proceedings/proceedings/sugi30/198-30.pdf.

Milliken, G. A., Johnson, D. E. (2004). Analysis of Messy data (Vol. 1). Designed experiments. New York: Chapman & Hall.

Molenberghs, G., & Verbeke, G. (2000). Linear models for longitudinal data. Springer.

Ohid Ullah, M., Hasan, M. A., Rahman, M. M., Chowdhury, A. H., & DAS, N.C., Jamal Uddin, M., Taj Uddin, M. . (2014). Obesity of primary school children: A cross-sectional study in Bangladesh. *International Journal of Scientific & Engineering Research, 5*(12), 263–270.

Tanner, J. M. (1951). Some notes on the reporting of growth data. *Human Biology, 23*(2), 93–159.

World Health Organization. (1995). Physical status: The use of and interpretation of anthropometry. Report of a WHO Expert Committee. http://www.who.int/childgrowth/publications/physical_status/en/.

World Health Organization. (2003). Obesity and Overweight. https://www.who.int/dietphysicalactivity/media/en/gsfs_obesity.pdf.

World Healthorganization. (2015). Commission on Ending childhood obesity. https://www.who.int/end-childhood-obesity/en/.

Statistical Analysis of Various Optimal Latin Hypercube Designs

A. R. M. Jalal Uddin Jamali, Md. Asadul Alam, and Abdul Aziz

Abstract ***Among several Design of Experiments (DoEs)***, Latin Hypercube Design (LHD) is one of the most frequently used methods in the field of physical experiments and in the field of computer simulations to find out the behavior of response surface of the surrogate model with respect to design points. A good experimental design should have three important characteristics, namely (i) non-collapsing, (ii) space-filling and (iii) orthogonal properties. Though inherently LHD preserves non-collapsing property, but randomly generated LHDs have poor space-filling in terms of minimum pair-wise distance. In order to ensure the last two properties in LHD, researchers are frequently looking for finding optimal LHD in the sense of space-filling and orthogonal properties. Moreover, researchers are frequently encountered the question, which distance measure is the best in the case of optimal designs? In the literature, several types of optimal LHDs are available such as Maximin LHD, Orthogonal LHD, Uniform LHD, etc. On the other hand, two distance measures namely Euclidean and Manhattan distance measures are used frequently to find optimal DoEs. But which one of the two distance measures is better, is still unknown. In this article, intensive statistical analysis has been carried out on numerical instances to explore the deep scenario of each optimal LHD. The main goal of this research is to find out a scenario of the well-known optimal designs from statistical point of view. From this elementary experimental study, it seems to us that in the sense of space-filling, Euclidean distance measure-based Maximin LHD is the best. But if one needs space-filling along with better orthogonal property, then multi-objective (Maximin with approximate orthogonal)-based optimal LHD is relatively better than Maximin LHD.

A. R. M. Jalal Uddin Jamali (✉)
Department of Mathematics, Khulna University of Engineering & Technology, Khulna, Bangladesh
e-mail: jamali@math.kuet.ac.bd

Md. Asadul Alam
Department of Mathematics, Daud Public College, Jessore, Bangladesh

A. Aziz
Department of CSE, Khulna University of Engineering & Technology, Khulna, Bangladesh

B. K. Sinha et al. (eds.), *Data Science and SDGs*,
https://doi.org/10.1007/978-981-16-1919-9_13

Keywords Experimental design · Space-filling · Non-collapsing · Orthogonal · Euclidean distance

1 Introduction

Computer simulation-based DoEs are used in a wide range of applications to learn about the effect of input variables x on a response of interest y. Fang et al. (2006) said, as simulation programs are usually deterministic so the output of a computer experiment is not subject to random variations, which makes the design of computer experiments different from that of physical experiments. Many simulation models involve several hundred factors or even more. It is desirable to avoid replicates when projecting the design onto a subset of factors. This is because a few out of the numerous factors in the system usually dominate the performance of the product. Thus, a good model can be fitted using only these few important factors. Therefore, when we project the design on to these factors, replication is not required. As it is recognized by several authors, the choice of the design points for computer experiments should at least fulfill two requirements namely non-collapsing and space-filling (Johnson et al. 1990; Morris and Mitchell 1995). In addition, in order to find out the individual effect of factors on response surface, researchers hunt DoEs with orthogonal property among the factors. LHD inherently preserves non-collapsing property, i.e. factors are evenly spread whenever projected on any coordinate. McKay et al. (1979) first introduced LHD. Later, authors (Morris and Mitchel 1995; Grosso et al. 2009) defined LHD which are a bit different. In a LHD, there are N points (design points) and each point has k distinct coordinates (parameters/factors). The points are placed in such a way that they are uniformly distributed when projected on every single coordinate (factor) axis. In a LHD, the range of each parameter/factor is normalized to the interval [1, N] (actually authors Grosso et al. 2009) considered the interval [0, N-1] but for the symmetry of other designs considered, here we have considered the range [1, N]). Mathematically, let us consider a set of N points in a uniform k-dimensional grid $\{1, 2, \ldots, N\}^k$:

$$X = \begin{pmatrix} \boldsymbol{X}_1 \\ \vdots \\ \boldsymbol{X}_N \end{pmatrix} = \begin{pmatrix} x_{11} & \ldots & x_{1k} \\ \vdots & \ldots & \vdots \\ x_{N1} & \ldots & x_{Nk} \end{pmatrix} \forall x_{ij} = \{1, 2, \ldots, N\}.$$

Then $\boldsymbol{X}$ is a LHD iff $x_{pj} \neq x_{qj} \, \forall \, j, p, q \in \{1, 2, \ldots, N\} \exists \, p \neq q$, i.e. each column has no duplicate entries. Note that the number of possible LHDs is huge: there are $(N!)^k$ possible LHDs.

It is known that randomly generated LHD has poor space-filling property. Moreover, it bears poor orthogonal property. Therefore, researchers search for optimal LHDs in the sense of space-filling, orthogonal, uniform etc., Jin et al. (2005), and Johnson et al. (1990) proposed Maximin (maximize the minimum inter-site pair-wise distance) experimental design but it was not LHD. They measured the space-filling on the basis of minimum inter-site pair-wise distance "D_1" among all pair-wise distances of design points. That is if "D_1" value of any LHD is relatively smaller then we say that the LHD is not good enough in the sense of space-filling. Morris and Mitchell (1995) proposed Φ_p optimal criterion and later Grosso et al. (2009) modified it for easy computation, which is given below:

$$\Phi_p(X) = \sum_{i=1}^{N} \sum_{j=i+1}^{N} \left[\frac{1}{d_{ij}^p} \right]^{\frac{1}{p}} \tag{1}$$

where $d_{ij} = d(x_i, x_j)$ be the distance between points x_i and x_j and p is a positive integer.

Audze and Eglais (1977) proposed a new optimal criterion that is similar to the potential energy function (Potential (U)) and which is defined as below:

$$Min\ U = \min \sum_{i=1}^{N} \sum_{j=i+1}^{N} \frac{1}{d_{ij}^2}. \tag{2}$$

It is known that *multicollinearity* (orthogonal property) measures the linear dependency among the factors of the design points. Multicollinearity may be measured by the partial pair-wise correlations among the factors. Several ways are available in the literature to measure the pair-wise correlations. Here, we consider the measure of average pair-wise correlations and maximum correlation, respectively, as below:

$$\rho^2 = \frac{\sum_{i=2}^{k} \sum_{j=1}^{i-1} \rho_{ij}^2}{k(k-1)/2} \tag{3}$$

and

$$\rho_{\max} = \max_{1 \le i, j \le k} \rho_{ij} \tag{4}$$

Here ρ_{ij} denotes the simple product–moment correlation between the factors i and j; k denotes the number of factors in the design considered. Butler (2001) found optimal LHD on the basis of this criterion. On the other hand, Joseph and Hung (2008) proposed the combination of the above two criteria. Moreover, Fang et al. (2000) defined a uniform design that allocates experimental points uniformly spread over the domain. It is noted that uniform design does not require orthogonal (multicollinearity) property. It considers projective uniformity over all sub-dimensions. But, in Fang et al. (2000), authors classified uniform design as space-filling design. A good literature review is available in Viana et al. (2014). It is worthwhile to mention here that Jamali et al. (2019) recently established some relations and bonds between Euclidian and Manhattan distance measures regarding LHD.

2 Experimental Result and Discussion

For the experimental study, we have considered a typical LHD of four factors ($k = 4$) with nine ($N = 9$) design points, which is optimized using several optimal criteria. The optimized LHDs are shown in Table 1. In Table 1, MLH-SA denotes Maximin LHD regarding Manhattan distance measure (L^1) obtained by Simulated Annealing (SA) approach (Husslage et al. 2011). OMLH-MSA denotes Orthogonal Maximin LHD regarding L^1 distance measure obtained by Modified Simulated Annealing (MSA) (Morris and Mitchell 1995), OLH-Y denotes Orthogonal LHD regarding L^1 distance measure obtained by Ye (1998), ULH-F denotes uniform LHD regarding L^1 distance measure obtained by Fang et al. (2000) and MLH-ILS indicates the Maximin LHD regarding Euclidean distance measure (L^2) obtained by Iterated Local Search (ILS) approach (Grosso et al. 2009).

At first, we have extracted all important statistical characteristics of each optimal design given in Table 1. Then all the findings are shown in Table 2. The symbols SD, CoV, β_1, β_2, Min Distance, Max Distance denote Standard Deviation, Coefficient of Variation (%), Moment Coefficient of Skewness, Moment Coefficient of Kurtosis ($\alpha_1 + 3$), minimum pair-wise inter-site distance among all design points of the LHD and maximum pair-wise inter-site distance among all design points of the LHD respectively. Besides, D_1^{L1} and D_1^{L2} denote the minimum pair-wise inter-site distance of design points measured in L^1 (Manhattan) distance measure and in L^2 (Euclidean) distance measure respectively. It is mentioned earlier that space-filling is defined on only D_1 value or Φ value. Similarly, orthogonal LHD is defined by multi-collinearity and so on. But the insight views of several optimized LHDs are still unknown. Here, we have analyzed the optimal LHDs statistically and on distance measures. In addition, we have compared optimized LHDs in those issues.

Table 1 Some optimal LHDs for $(k,N) = (4,9)$

Method	LH-SA	OMLH-MSA	OLH-Y	ULH- F	MLH-ILS
Optimal design matrix	1 3 3 4	1 5 3 3	1 2 6 3	4 1 7 5	1 5 8 4
	2 5 8 8	2 2 5 8	2 9 7 6	1 3 4 3	2 7 4 9
	3 8 6 2	3 9 7 5	3 4 2 9	9 9 5 4	3 2 1 6
	4 7 1 6	4 3 8 1	4 7 1 2	6 6 6 9	4 8 3 3
	5 2 9 3	5 7 1 7	5 5 5 5	5 7 2 1	5 1 5 1
	6 9 5 9	6 6 9 9	6 3 9 8	2 8 8 7	6 3 7 8
	7 1 4 7	7 1 2 4	7 6 8 1	3 5 1 6	7 6 9 2
	8 4 2 1	8 8 4 2	8 1 3 4	8 2 3 8	8 9 6 7
	9 6 7 5	9 4 6 6	9 8 4 7	7 4 9 2	9 4 2 5

Source Collected from the following papers: Morris and Mitchell (1995), Ye (1998), Fang et al. (2000) and Grosso et al. (2009)

Table 2 Statistical analysis among several optimal LHDs for $(k, N) = (4, 9)$

Method	MLH-SA	OMLH-MSA	OLH-Y	ULH-F	MLH-ILS
Objective function	Maximin	Maximin + ρ	ρ	Uniform	Maximin
Dist. used in the method	L^1	L^1	L^1	L^1	L^2
No. of distances (n)	36	36	36	36	36
Sum of distances	2160	2160	2160	2160	2160
Average	60	60	60	60	60
SD	21.7894	**21.19355**	24.4949	21.65641	22.59794
CoV	36.31566	**35.32259**	40.82483	36.09401	37.66323
β_1	1.237749	**0.989075**	1.278651	0.148856	1.332428
$\beta_2(\alpha_1 + 3)$	0.760507	***0.093569***	1.918449	**–0.93998**	0.347584
Min Distance (L^2)	33	31	30	26	**42**
Max distance (L^2)	122	**109**	120	102	112
Range (L^2)	89	78	90	76	**70**
ρ	0.088889	0.041667	**0**	0.061111	0.133333
ρ_{Max}	0.216667	0.116667	**0**	0.15	0.233333
Φ_p^{L2}	0.183966	0.187146	0.202608	0.207282	**0.17497**
Potential (U)	0.667812	0.66931	0.7	0.699	**0.666661**
D_1^{L1}	**11**	**11**	10	10	10
D_1^{L2}	33	31	30	26	**42**

Source Authors' own calculation by using programs

We have also detected all pair-wise inter-site Euclidean distances (actually square of Euclidean distances) "$D_{i,j}^2$" values for all pair of design points for each optimal LHDs. Now for each optimal LHD, the "$D_{i,j}^2$" are plotted corresponding to each pair of design points (I = {1,2,..., 36}), which are shown in Fig. 1. It is noted that since we have taken 9 design points, we have obtained $[(N(N-1)/2] = 36$ number of pair-wise distances. It is worthwhile to mention here that the sum of square of all pair-wise (Euclidian) distances, $\left(d\left(x_i, x_j\right)\right)^2$, of a fixed LHD is identical (=2160) and so average of the square distances of all LHDs (for $k = 4$, $N = 9$) are also identical. It is noted that there are $(9!)^4$ such LHDs and for each such LHD the value of $\sum \left(d\left(x_i, x_j\right)\right)^2$ is equal to 2160.

Now, we will discuss the comparison between two Maximin LHDs namely MLH-SA and MLH-ILS in which both are optimized by using same single optimal criterion called Maximin optimal criterion. But distance measures namely L^1 and L^2 are used in MLH-SA and MLH-ILS, respectively, which are different. In addition, different optimal heuristic algorithms namely SA and ILS approaches are applied to optimize the MLS-SA and MLH-ILS designs, respectively. It is observed in both MLH-SA and MLH-ILS that the values of SD, CoV, β_1, β_2, ρ, ρ_{Max}, Φ_p^{L2} and Potential (U) are not significantly different. It is also observed that the difference of D_1^{L1} of the

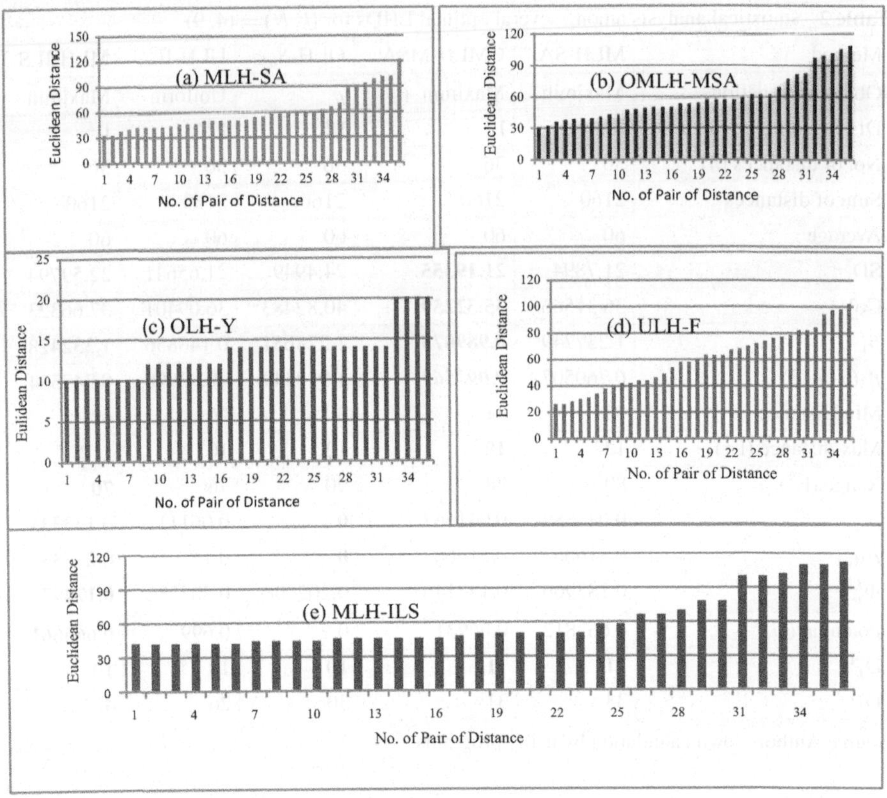

Fig. 1 Graphical representation of square of Euclidean distances corresponding to the pair of design points for each optimal LHDs where $(k, N) = (4, 9)$. *Source* Created by the authors

two DoEs is not significant, though MLH-SA searched optimal D_1^{L1} value whereas MLH-ILS explored for D_1^{L2} value rather than D_1^{L1}.

Now we will talk about the comparison between Maximin LHD and Uniform LHD mainly MLH-ILS and ULH-F. It is shown in Table 2 that uniform optimality criterion-based LHD is flatter and un-skewed compared to all other LHDs considered here. Moreover, the coefficient of correlation of ULH-F is relatively smaller than that of MLH-ILS. It is also observed that SD, CoV, Max Distance and Range values are not significantly different from the two optimal LHDs namely MLH-ILS and ULH-F. It is also clear from Table 2 that the D_1^{L1} values of both DoEs are identical. On the other hand, for Φ_p^{L2}, Potential (U) and D_1^{L2} values, MLH-ILS design is significantly better than those of ULH-F design. It is worthwhile to mention here that the D_1^{L2} value of ULH-F is worst compared to other optimal designs considered here.

We will also discuss orthogonal-based DoEs mainly of OLH- Y. In addition, comparison will be drawn with Maximin-based LHD mainly MLH-ILS. Since the initial criterion of OLH-Y is $\rho = 0$, so obviously the factors of the DoE are all uncorrelated. It is observed in Table 2 that the SD, CoV, β_1 and D_1^{L1} values of OLH-Y are not significantly different with that of MLH-ILS. But it is noticed that for β_2, Max Distance, Range, Φ_p^{L2}, Potential (U) and D_1^{L2} values, OLH-Y is significantly worse compared with the values of MLH-ILS. Moreover, though in OLH-Y, L^1 distance measure is used whereas, in MLH-ILS, L^2 distance measure is used but both D_1^{L1} values are identical. On the other hand, D_1^{L2} value of MLH-ILS is significantly better compared with not only OLH-Y design but also all other designs considered here.

Now we will discuss the properties of OMLH-MSA in which multi-objective function namely Maximin and multi-collinearity was considered. It is observed that the SD, the CoV, the β_1 and Max Distance values of OMLH-MSA are the best compared with all other optimal designs. But these values are not significantly different from those values of MLH-ILS.

It is observed in Table 2 that the average correlation as well as maximum correlation of OMLH-MSA is negligible and smaller than those values of MLH-ILS. But it is also noticed in Table 2 that the average correlation as well as maximum correlation of MLH-ILS is also not significantly larger. In addition, it is also observed in Table 2 that D_1^{L1} value of OMLH-MSA is not significantly smaller than that of other optimal LHDs. On the other hand, D_1^{L2}, Φ_p^{L2} and potential (U) values of MLH-ILS are the best compared with all other optimal LHDs. Moreover, these values are significantly better compared with OMLH-MSA. It is also observed in Fig. 1 that according to space-filling criterion MLH-ILS is relatively looking more homogeneous over the design space compared with other designs.

Now we will discuss the two distance measures considered here namely Euclidean and Manhattan. It is noticed in Table 2 that except MLH-ILS all other designs are optimized by using Manhattan distance measure whereas MLH-ILS is optimized by considering Euclidean distance measure. It is observed in Table 2 that D_1^{L1} value of anyone LHD compared with the remaining other LHDs is not significantly better. On the other hand, the D_1^{L2} value of MLH-ILS is the best and significantly larger than any other one.

3 Conclusion

When LHD is considered as an experimental design, the design inherently preserves good non-collapsing property. On the other hand, when Maximin optimal criterion is considered, then the DoE should have good space-filling property. Again, when we use multicollinearity as an optimal criterion, then the factors of the DoE are uncorrelated or approximately uncorrelated. Similarly, whenever the uniform optimal design is considered then design becomes flatter. Besides optimal criteria, another question is frequently raised; which distance measure is relatively better in case of an optimal design? In this elementary study, the main objective is to find out the schematic view of several well-known optimal designs from the statistical point of view. A rigorous statistical analysis has been carried out over some well-known optimal LHDs, which are optimized with different optimal criteria. From this elementary analysis, it seems to us that, if we prefer a DoE containing all three properties (good) then multi-objective-based LHD namely OMLH-MSA is relatively better. But if one considers only space-filling LHD, then MLH-ILS is the best in which Euclidean distance measure is considered. Similarly, if one needs the design with uncorrelated factors along with non-collapsing property then it is obvious that OLH-Y is the best. In addition, it seems to us that L^2 is relatively better to find out Maximin LHD. Though multi-objective-based optimal LHD is relatively better, it is obviously time-consuming. Moreover, in that optimal DoE, none of the individual optimal criterion (of multi-objective function) is found good enough. It is worthwhile to mention here that ILS is preferable to find Maximin LHD with cheaper computational time. Therefore, according to us, one may first find out an optimal LHD by considering Maximin optimal criterion with ILS approach to detect the important factors on response surface by using L^2 distance measure. Then orthogonal optimal criteria may be applied on the reduced DoE (eliminated unimportant factors) to get an uncorrelated-based DoE for further experiments. But it is worthwhile to mention here that, in this experimental research, only some optimal LHDs of $(k, N) = (4, 9)$ are considered. At the same time, we would like to note here that in the existing literature only four optimal LHDs of $(k, N) = (4, 9)$ viz MLH-SA, OMLH-MSA, OLH-Y, ULH-F, MLH-ILS are available, which means, some other instances that are optimized with all these five approaches are not yet available in the literature. So, further experiments (with different instances) should be carried out to draw a more concrete conclusion.

Acknowledgements This research work is supported by the Special Grant of Ministry of Science and Technology and Khulna University of Engineering and Technology, Khulna, Bangladesh.

References

Audze, P., & Eglais, V. (1977). New approach for planning out of experiments. *Problems of Dynamics and Strength, 35*, 104–107.

Butler, N. A. (2001). Optimal and orthogonal Latin hypercube designs for computer experiments. *Biometrika, 88*(3), 847–857.

Fang, K. T., Lin, D. K. J., Winker, P., & Zhang, Y. (2000). Uniform design: Theory and application. *Technimetrics, 42*(3), 237–248.

Fang, K. T., Li, R., & Sudjianto, A. (2006). *Design and modeling for computer experiments*. CRC Press.

Felipe, A. C. V., Simpson, T. W., Balabanov, V., & Toropov, V. (2014). Metamodeling in multi-disciplinary design optimization: How far have we really come? *AIAA (American Institute of Aeronautics and Astronautics) Journal, 52*(4), 670–690.

Grosso, A., Jamali, A. R. J. U., & Locatelli, M. (2009). Finding maximin latin hypercube designs by I. L. S. Heuristics . *European Journal of Operation Research, 197*, 541–547.

Husslage, B. G. M., Rennen, G., van Dam, E. R., & den Hertog, D. (2011). Space-filling Latin hypercube designs for computer experiments. *Journal of Optimization and Engineering, 12*, 611–630. https://doi.org/10.1007/s11081-010-9129-8

Jamali, A. R. M. J. U., Alam, Md., & A. . (2019). Approximate relations between Manhattan and Euclidean distance regarding Latin hypercube experimental design . *Journal of Physics: Conference Series IOP, 1366*, 012–030. https://doi.org/10.1088/1742-6596/1366/1/012030

Jin, R., Chen, W., & Sudjianto, A. (2005). An efficient algorithm for constructing optimal design of computer experiments. *Journal of Statistical Planning and Inference, 134*(1), 268–287.

Johnson, M. E., Moore, L. M., & Ylvisaker, D. (1990). Minimax and maximin distance designs. *Journal of Statistical Planning and Inference, 26*, 131–148.

Joseph, V. R., & Hung, Y. (2008). Orthogonal-maximin latin hypercube designs. *Statistica Sinica, 18*, 171–186.

McKay, M. D., Beckman, R. J., & Conover, W. J. (1979). A comparison of three methods for selecting values of input variables in the analysis of output from a computer code. *Techno Metrics, 21*, 239–245.

Morris, M. D., & Mitchell, T. J. (1995). Exploratory designs for computer experiments. *Journal of Statistical Planning and Inference, 43*, 381–402.

Ye, K. Q. (1998). Orthogonal column Latin hypercube and their application in computer experiments. *Journal of the American Statistical Association, 3*, 1430–1439.

Erlang Loss Formulas: An Elementary Derivation

Jyotirmoy Sarkar

Abstract The celebrated Erlang loss formulas, which express the probability that exactly j of c available channels/servers are busy serving customers, were discovered about 100 years ago. Today we ask: "What is the simplest proof of these formulas?" As an alternative to more advanced methods, we derive the Erlang loss formulas using (1) an intuitive limit theorem of an alternating renewal process and (2) recursive relations that are solved using mathematical induction. Thus, we make the Erlang loss formulas comprehensible to beginning college mathematics students. We illustrate decision making in some practical problems using these formulas and other quantities derived from them.

Keywords Alternating renewal process · Cycle time · Ergodicity · Queuing system · Renewal time · Semi-Markov process

MSC Code: Primary 60J28 · 60K20 · Secondary 62-01

1 Introduction

The *Erlang loss model*, also known as the $M/M/c/c$ queuing system, describes the stochastic behavior of the number of customers receiving service at a service center Suppose that customers arrive according to a Poisson process with rate λ; that is, the inter-arrival times between successive customers are independent and identically distributed (IID) exponential(λ) variables with mean $1/\lambda$. The service station has c identical channels/servers with *priority ordering* in providing service, determined by their identification numbers $1, 2, 3, \ldots, c$. On arrival, a customer chooses the channel with the lowest ID that is free. If all channels are busy, the customer leaves the service station, and is lost to the system. Since no customer waits for a delayed

---

J. Sarkar (✉)
Indiana University–Purdue University Indianapolis, Indianapolis, IN 46202, USA
e-mail: jsarkar@iupui.edu

B. K. Sinha et al. (eds.), *Data Science and SDGs*,
https://doi.org/10.1007/978-981-16-1919-9_14

commencement of service, no queues are actually formed in this queuing system! The service times at each channel/server are IID exponentila(μ).

The Erlang loss model is well-known in the literature. The model arises in statistical reliability theory in the context of availability of multi-component systems. See, for example, Mann et al. (1974), Rausand and Høyland (2004) and Sarkar and Li (2006). The model is applicable in traditional services such as customers at vending machines, photocopy machines, bank tellers, car ride services, phone lines and repairing facilities, and also in new services such as computer server access, smart phone network access, etc. Oftentimes, resource limitations dictate the number of channels or servers available at the service station. Also, availability of competing service providers motivate potential customers to leave if they find none of the channels/servers free at the time of their need.

In an Erlang loss model, different questions are of interest to different parties: (1) What proportion of customers is lost to the $M/M/c/c$ system? (2) What proportion of times Channel d ($1 \le d \le c$) is busy? (3) How much is the expected free time for Channel d? The service provider wants to know how many channels to install and operate. Each server wants to know what proportion of time she will be busy, what proportion of customers she will serve, and between successive services for how long she is likely to be free. The customer wants to know the probability that he will receive service at this center; and if served, which server is likely to provide the service.

To answer these questions, the stochastic behavior of the $M/M/c/c$ queuing system (henceforth called the $M/M/c/c$ process) is described by a continuous-time stochastic process (CTSP) on the discrete state space $\{0, 1, 2, \ldots, c\}$, where State j means that exactly j (we do not need to specify which j) among Channels $1, 2, \ldots, c$ are busy. The proportion of time exactly j channels ($0 \le j \le c$) are busy, is given by

$$\theta_j = \frac{\rho^j/j!}{A_c(\rho)} \quad \text{for } j = 0, 1, \ldots, c \tag{1}$$

where

$$A_c(\rho) \equiv \sum_{i=0}^{c} \frac{\rho^i}{i!} = 1 + \rho + \frac{\rho^2}{2!} + \frac{\rho^3}{3!} + \ldots + \frac{\rho^c}{c!} \tag{2}$$

Thus, the θ_j's are truncated versions of probabilities of a Poision(ρ) random variable. Expressions in (1) are called the *Erlang loss formulas*.

It has been exactly 100 years since Agner Krarup Erlang discovered the celebrated loss formulas that bear his name. We quote from O'Connor and Robertson: "In 1917 he (Agner Erlang) published *Solution of some problems in the theory of probability of significance in automatic telephone exchanges* in which he gave a formula for loss and waiting time which was soon used by telephone companies in many countries including the British Post Office."

In the literature, the Erlang loss formulas are derived using one of two advanced methods: (i) Use the ergodic theorem of a time-homogeneous, irreducible, non-lattice continuous-time Markov process. See Medhi (1982), for example. The proof of the theorem is based on Kolmogorov's forward (differential) equations, which is a deep result whose proof is delegated to advanced graduate courses in stochastic processes. (ii) Use the ergodic theorem of a semi-Markov process. See Ross (1996), for example. This proof uses the strong law of large numbers, and is typically presented in an introductory graduate course in stochastic processes. Both of these approaches leave the general readership of undergraduate students or non-math/stat specialists substantially perplexed. Therefore, we are motivated to offer them a simpler derivation of the Erlang loss formulas. We do so using an intuitive alternating renewal theorem and several recursive relations that are solved by mathematical induction, very much in the same spirit as in Sarkar (2006), which studies the random walk on the vertices of a polygon, only now with more complicated transition probabilities.

The rest of the paper is organized as follows: In Sect. 2 we present the details of our simpler derivation. In Sect. 3, we answer Questions 1–3, and some other related questions. In Sect. 4, we illustrate decision making by evaluating these formulas in two applied problems. Section 5 concludes the paper with some remarks.

2 An Elementary Derivation

We utilize the following intuitively appealing limit theorem.

Theorem 2.1 (Limit theorem of an alternating renewal process) *Suppose that a CTSP that is alternating between "on" and "off" states. Let the sequence of (on time, off time) be IID with a finite mean vector* $(\nu_{\text{on}}, \nu_{\text{off}})$. *Then the proportion of "on" time within the time interval* $(0, t)$ *converges to* $\nu_{\text{on}}/(\nu_{\text{on}} + \nu_{\text{off}})$, *as* $t \to \infty$.

The proof of Theorem 2.1 is based on the elementary renewal theorem, and it can be found in many textbooks. See, for example, Medhi (1982), Rausand and Høyland (2004) or Ross (1996).

To apply Theorem 2.1 to our $M/M/c/c$ process, let us specify the contextual meanings of *on time, off time, renewal epoch* and *cycle time*: The CTSP is undergoing an alternating sequence of "all c channels busy" (*on time*) and "not all c channels busy" (*off time*). The *renewal epoch* of this CTSP is the instant when the process begins an "on time" (that is, it enters State c). The duration between successive renewal epochs is called the *cycle time*. The expected cycle time is $\nu_{\text{on}} + \nu_{\text{off}}$.

When the CTSP is in State c, any new arrival is lost to the system. Therefore, the sojourn time in State c, being the smallest among c IID exponential(μ) variables corresponding to the c customers being served, is exponential($c\mu$). Therefore, the expected sojourn time in State c is $\nu_{\text{on}} = 1/(c\mu)$. From state c the CTSP surely enters state $(c - 1)$, and "off time" commences. But how much is ν_{off}, before the CTSP re-enters state c? We answer this question eventually in Proposition 2.1 below. But first let us track the $M/M/c/c$ process as it evolves from State $(c - 1)$.

From state $(c-1)$, either the CTSP returns to state c directly, or it goes on to visit one or more lower-numbered states before making its way up again. In fact, if the CTSP visits a State j (for $0 < j < c$), then it remains there for a duration given by the minimum of j IID exponential(μ) variables (corresponding to j customers being served) and another independent exponential(λ) variable (corresponding to the arrival of a new customer). Hence, the sojourn time in state j has an expected duration $1/(\lambda + j\mu)$, before the CTSP moves to one of the neighboring states $(j+1)$ or $(j-1)$ with respective probabilities in the ratio $\lambda : j\mu$. Similarly, if the CTSP ever visits state 0, it stays there until the arrival of a new customer; that is, it stays there for an expected duration $1/\lambda$ before it surely moves to State 1. Therefore, focusing only on the transition epochs of the CTSP, we identify an embedded discrete time stochastic process (DTSP). It has state space $\{0, 1, 2, \ldots, c\}$ and transition probabilities

$$\begin{cases} p(0,1) = 1 \\ p(j, j+1) = \frac{\lambda}{\lambda + j\mu} & \text{for } j = 1, 2, \ldots, (c-1) \\ p(j, j-1) = \frac{j\mu}{\lambda + j\mu} & \text{for } j = 1, 2, \ldots, (c-1) \\ p(c, c-1) = 1 \end{cases} \tag{3}$$

Thus, the CTSP is a semi-Markov process.

We need a notation. For $0 \le h, i, j \le c$, let $\tau_j(h : i)$ denote the expected time the CTSP spends in state j while going from State h to State i, and let $\tau_+(h : i)$ denote the expected total time the CTSP spends while going from State h to State i. First, in Lemma 2.1, we give the expected time needed to go "one step up" from state j to state $(j+1)$. This idea has been used before, for example in Abate and Whitt (1998). In particular, $\nu_{\text{off}} = \tau_+(c-1, c)$. Then in Proposition 2.1, we evaluate the expected cycle time $\tau_+(c : c)$; and finally in view of Theorem 2.1, we find θ_c.

Lemma 2.1 *In the $M/M/c/c$ process, the expected time to go from j to $(j+1)$ is*

$$\tau_+(j : j+1) = \frac{1}{\lambda} \frac{A_j(\rho)}{\rho^j/j!} \tag{4}$$

Proof of Lemma 2.1 Note that $\tau_+(0 : 1) = 1/\lambda$. For $j \ge 1$, after a sojourn in State j for an expected duration of $1/(\lambda + j\mu)$, either the CTSP goes to State $(j+1)$ directly, or it goes there later on after first dropping down to State $(j-1)$ and then making its way up. Hence, using (3), we have

$$\begin{aligned} \tau_+(j : j+1) &= \frac{1}{\lambda + j\mu} + \frac{j\mu}{\lambda + j\mu} \tau_+(j-1 : j+1) \\ &= \frac{1}{\lambda + j\mu} + \frac{j\mu}{\lambda + j\mu} \{\tau_+(j-1 : j) + \tau_+(j : j+1)\} \end{aligned}$$

since a passage from $(j-1)$ to $(j+1)$ always goes through j. Or equivalently,

$$\tau_+(j : j+1) = \frac{1}{\lambda}\{1 + j\mu\,\tau_+(j-1 : j)\} \tag{5}$$

To solve the recursive relation (5), we use mathematical induction on $j = 0, 1, 2, \ldots$. Clearly, the lemma holds for $j = 0$, since $\tau_+(0 : 1) = 1/\lambda$ and $A_0(\rho) = 1$. Assume that the lemma holds for some $(j-1) \geq 0$. By the induction hypothesis, from (5) we have

$$\tau_+(j : j+1) = \frac{1}{\lambda}\left\{1 + j\mu\,\frac{1}{\lambda}\,\frac{A_{j-1}(\rho)}{\rho^{j-1}/(j-1)!}\right\} = \frac{1}{\lambda}\left\{1 + \frac{A_{j-1}(\rho)}{\rho^j/j!}\right\} = \frac{1}{\lambda}\left\{\frac{A_j(\rho)}{\rho^j/j!}\right\}$$

So the lemma holds for j, and the proof is complete. Q.E.D

Proposition 2.1 *In the $M/M/c/c$ process, the expected cycle time is*

$$\tau_+(c : c) = \nu_{\text{on}} + \nu_{\text{off}} = \frac{1}{c\,\mu}\left\{1 + \frac{A_{c-1}(\rho)}{\rho^c/c!}\right\} = \frac{1}{c\,\mu}\,\frac{A_c(\rho)}{\rho^c/c!} \tag{6}$$

and the proportion of time all c channels are busy is $[\rho^c/c!]\,/A_c(\rho)$.

Proof of Proposition 2.1 Clearly, $\tau_+(c : c) = \tau_c(c : c-1) + \tau_+(c-1 : c) = \nu_{\text{on}} + \nu_{\text{off}}$. We already proved that $\nu_{\text{on}} = \tau_c(c : c-1) = 1/(c\mu)$. Also, specializing (4) to $j = (c-1)$, we get

$$\nu_{\text{off}} = \tau_+(c-1 : c) = \frac{1}{\lambda}\,\frac{A_{c-1}(\rho)}{\rho^{c-1}/(c-1)!} = \frac{1}{c\mu}\,\frac{A_{c-1}(\rho)}{\rho^c/c!}$$

Adding these two components, we prove (6). Also, the ratio of $\tau_c(c : c-1)$ to $\tau_+(c : c)$, in view of Theorem 2.1, gives the proportion of time all c channels are busy. Q.E.D

Proposition 2.1 gives the proportion of time the CTSP spends in State c. The next proposition gives the expected time the CTSP spends in states $0, 1, \ldots, (c-1)$ within a cycle time of the $M/M/c/c$ process, and hence establishes (1).

Proposition 2.2 *In the notation described above, for* $0 \leq j < c$

$$\tau_j(c : c) = \frac{(c-1)!/j!}{\lambda\,\rho^{c-1-j}} \tag{7}$$

Furthermore, the proportion of time the CTSP spends in State j is as given in (1).

The proof of Proposition 2.2 is based on two lemmas, which require two new notation. Let $q_j(h : c)$ denote the probability that the embedded DTSP visits State j on its way from State h to State c. In particular, $q_j(h : h) = \delta_{h,j}$, the Kronecker's delta function taking value 1 when $j = h$, and 0 otherwise. Also, let $B_0(\rho) = 1$, and for $0 \leq j < (c-1)$, let

$$B_{c-1-j}(\rho) = 1 + \frac{\rho}{c-1} + \frac{\rho^2}{(c-1)(c-2)} + \ldots + \frac{\rho^{c-1-j}}{(c-1)!/j!} \tag{8}$$

Lemma 2.2 expresses $\tau_j(c:c)$ in terms of $q_j(c-1:c)$ and $1-q_j(j+1:c)$, each of which is then expressed in terms of $B_{c-1-j}(\rho)$ in Lemma 2.3.

Lemma 2.2 *In the notation described above, for* $0 \leq j < c$

$$\tau_j(c:c) = q_j(c-1:c)\ \left[\lambda\left\{1-q_j(j+1:c)\right\}\right]^{-1} \tag{9}$$

Proof of Lemma 2.2 Note that for any $0 \leq j < c$, we have

$$\tau_j(c:c) = \tau_j(c-1:c) = q_j(c-1:c)\ \tau_j(j:c) \tag{10}$$

We next evaluate the second factor on the right hand side of (10). Indeed, using (3), for $0 \leq j < c$, we have the recursive relation

$$\tau_j(j:c) = \frac{1}{\lambda+j\mu} + \frac{j\mu}{\lambda+j\mu}\tau_j(j-1:c) + \frac{\lambda}{\lambda+j\mu}\tau_j(j+1:c)$$

Note that a transition from $(j-1)$ to c necessarily passes through j, but a transition from $(j+1)$ to c may drop down to j only with probability $q_j(j+1:c)$. Each time it visits State j, the CTSP spends an expected time $\tau_j(j:c)$ in state j. Hence, we have

$$\tau_j(j:c) = \frac{1}{\lambda+j\mu} + \frac{j\mu}{\lambda+j\mu}\tau_j(j:c) + \frac{\lambda}{\lambda+j\mu} q_j(j+1:c)\ \tau_j(j:c)$$

which simplifies to

$$\tau_j(j:c) = \left[\lambda\left\{1-q_j(j+1:c)\right\}\right]^{-1} \tag{11}$$

Therefore, substituting (11) in (10), we complete the proof of the lemma. Q.E.D

Lemma 2.3 *In the notation described above, for* $0 \leq j < c$

$$q_j(c-1:c) = [B_{c-1-j}(\rho)]^{-1} \tag{12}$$

$$1-q_j(j+1:c) = \frac{\rho^{c-1-j}}{(c-1)!/j!}\ [B_{c-1-j}(\rho)]^{-1} \tag{13}$$

Proof of Lemma 2.3 We will solve the paired equations ((12), (13)) using mathematical induction *backwards* starting from $j = (c-1)$. Clearly, (12) and (13) hold for $j = (c-1)$, since $q_{c-1}(c-1:c) = 1 = [B_0(\rho)]^{-1}$ and $q_{c-1}(c:c) = 0$, implying that $1-q_{c-1}(c:c) = 1 = [B_0(\rho)]^{-1}$. Assume that (12) and (13) hold for some $(j+1) \leq (c-1)$. We will show that (12) and (13) also hold for j. Note that, since after each transition, the CTSP moves only to a neighboring state, we have

$$q_j(c-1:c) = q_{j+1}(c-1:c)\, q_j(j+1:c) \tag{14}$$

The second factor on the right hand side of (14) is the one-step-drop-down probability of reaching j while the CTSP goes from $(j+1)$ to c. Such a drop can happen either at the very next transition; or it can happen later on following a one-step-up transition to $(j+2)$, and then eventually a one-step-down transition back to $(j+1)$ and another one-step-down transition to j before reaching c. Therefore, using (3), we have

$$q_j(j+1:c) = \frac{(j+1)\mu}{\lambda+(j+1)\mu} + \frac{\lambda}{\lambda+(j+1)\mu}\, q_{j+1}(j+2:c)\, q_j(j+1:c)$$

which simplifies to

$$\begin{aligned} q_j(j+1:c) &= \frac{(j+1)\mu}{\lambda+(j+1)\mu-\lambda\; q_{j+1}(j+2:c)} \\ &= \left[1+\frac{\rho}{j+1}\left\{1-q_{j+1}(j+2:c)\right\}\right]^{-1} \end{aligned} \tag{15}$$

Substituting (15) in (14), we have

$$q_j(c-1:c) = q_{j+1}(c-1:c)\left[1+\frac{\rho}{j+1}\left\{1-q_{j+1}(j+2:c)\right\}\right]^{-1}$$

to which we now apply the backward induction hypotheses that (12) and (13) hold for $(j+1)$, in order to get

$$\begin{aligned} q_j(c-1:c) &= \left\{B_{c-2-j}(\rho)\right\}^{-1}\left[1+\frac{\rho}{j+1}\,\frac{\rho^{c-2-j}}{(c-1)!/(j+1)!}\left\{B_{c-2-j}(\rho)\right\}^{-1}\right]^{-1} \\ &= \left[B_{c-2-j}(\rho)+\frac{\rho^{c-1-j}}{(c-1)!/j!}\right]^{-1} = \left[B_{c-1-j}(\rho)\right]^{-1} \end{aligned}$$

This establishes expression (12) for j. Next, from (14), we also have

$$q_j(j+1:c) = \frac{q_j(c-1:c)}{q_{j+1}(c-1:c)} = \frac{B_{c-2-j}(\rho)}{B_{c-1-j}(\rho)}$$

whence, in view of (8), we see that (13) holds for j. Thus, by mathematical induction, (12) and (13) hold for all $j = (c-1), (c-2), \ldots, 0$, completing the proof. Q.E.D

Proof of Proposition 2.2 Substituting (12) and (13) in (9), we get (7). Thereafter, taking the ratio of (7) to (6), we establish the Erlang loss formulas given in (1). Q.E.D

3 Answering Questions 1–3

To answer Question 1 we simply compute θ_c, the proportion of time all c channels are busy, since that is exactly the condition under which a new arrival overflows out of the service center. Therefore, as given in (1), the proportion of customers lost to the service center is given by $\omega_c = \theta_c = [\rho^c/c!]/A_c(\rho)$. Consequently, the proportion of customers served by the $M/M/c/c$ queuing system is $(1-\omega_c)$.

Thus, a proportion $(1-\omega_c)$ of customers are served by all c channels combined. A natural follow-up question is: What proportion of customers are served by each of Channels 1 through c? To answer this follow-up question, note that everything we have proved so far about the $M/M/c/c$ queuing system, also holds for the $M/M/d/d$ sub-system (consisting of Channels 1 through d), for $1 \le d \le c$. Henceforth, to avoid confusion, when we refer to the $M/M/d/d$ sub-system, the proportion of time exactly j among Channels 1 through d are busy will be written as $\theta_j(d) = [\rho^j/j!]/A_d(\rho)$. In particular, the proportion of time all Channels 1 through d are busy is

$$\omega_d = \theta_d(d) = \frac{\rho^d/d!}{A_d(\rho)} \tag{16}$$

A word of caution is warranted: From (16), we have $w_d = \theta_d(d)$. But $w_d \neq \theta_d(c)$, since the denominator of (16) is different from that in (1). In particular, $\omega_1 = \rho/(1+\rho)$, $\omega_2 = (\rho^2/2)/(1+\rho+\rho^2/2)$. Let us write $\omega_0 = 1$, which is consistent with (16), since $A_0(\rho) = 1$.

Now ϕ_d, the proportion of customers who are served by Channel d, is given by the difference between the proportion of customers lost to the $M/M/(d-1)/(d-1)$ sub-system (consisting of Channels 1 through $(d-1)$) and the proportion of customers lost to the $M/M/d/d$ sub-system. That is,

$$\phi_d = \omega_{d-1} - \omega_d = \theta_{d-1}(d-1) - \theta_d(d) = \frac{\rho^{d-1}/(d-1)!}{A_{d-1}(\rho)} - \frac{\rho^d/d!}{A_d(\rho)} \tag{17}$$

In particular, $\phi_1 = (1+\rho)^{-1}$, $\phi_2 = (1+\rho)^{-1}(\rho+\rho^2/2)/(1+\rho+\rho^2/2) < \phi_1$. We leave it to the reader to establish that $\phi_1 > \phi_2 > \ldots > \phi_c$; that is, a channel with a lower ID serves more customers than a channel with a higher ID, as anticipated from the priority ordering of the channels/servers. Alternatively, the reader may refer to Messerli (1972).

Note that ϕ_d, given in (17), is also the limiting probability that a randomly arriving customer will be served by Channel d. Furthermore, (17) can be re-written as

$$\phi_d = \frac{\rho^{d-1}/(d-1)!}{A_{d-1}(\rho)\,A_d(\rho)}\left\{A_d(\rho) - \frac{\rho}{d}A_{d-1}(\rho)\right\} = \frac{\omega_{d-1}}{A_d(\rho)}\sum_{i=0}^{d}\left(1-\frac{i}{d}\right)\frac{\rho^i}{i!}$$

which is a positive quantity. This, in view of (17), proves that $\omega_1 > \omega_2 > \ldots > \omega_c$. Of course, this inequality can be easily anticipated, since the overflow out of a larger system is smaller than the overflow out of a sub-system.

To answer Question 2, we have to find ξ_d, the proportion of time Channel d is busy ($1 \leq d \leq c$). Clearly, $\xi_1 = \theta_1(1) = \omega_1 = \rho/(1+\rho)$. For $d > 1$, let us write two equivalent expressions for $E[N_d]$, the expected number of channels (among Channels 1 through d) that are busy under the stationary distribution for the $M/M/d/d$ sub-system. The first expression comes from the definition of expectation and the Erlang loss formulas (1) applied tor the $M/M/d/d$ sub-system.

$$E[N_d] = \sum_{i=0}^{d} i\,\theta_i(d) = \sum_{i=1}^{d} i\,\theta_i(d) = \sum_{i=1}^{d} \frac{\rho^i/(i-1)!}{A_d(\rho)} = \frac{\rho\,A_{d-1}(\rho)}{A_d(\rho)} = \rho\,[1-\omega_d] \tag{18}$$

The second equivalent expression for $E[N_d]$ uses $N_d = \sum_{i=1}^{d} I(\text{Channel } i \text{ is busy})$, where $I(\cdot)$ denotes the indicator function. Hence,

$$E[N_d] = \sum_{i=1}^{d} P\{\text{Channel } i \text{ is busy}\} = \sum_{i=1}^{d} \xi_i \tag{19}$$

Equating the right hand sides of (18) and (19), we obtain

$$\xi_d = E[N_d] - E[N_{d-1}] = \rho\,[\omega_{d-1} - \omega_d] = \rho\,\phi_d \tag{20}$$

The last equality in (20) follows from (17). Again (20) agrees with our intuition that the time spent serving customers must be proportional to the number of customers served, since the service times are IID. Furthermore, note that $0 < \xi_d = \rho\,\phi_d < \rho\,\phi_1 = \rho/(1+\rho) < 1$, which ensures that ξ_d is a genuine proportion. Furthermore, note that $\xi_1 > \xi_2 > \ldots > \xi_c$, since $\phi_1 > \phi_2 > \ldots > \phi_c$.

Finally, to answer Question 3, we consider the alternating sequence of "busy" and "free" times of Channel d alone. We can define the cycle time of Channel d, σ_d say, as the sum of a busy time and the next free time of Channel d. We know that the expected busy time of Channel d (actually, that of any channel) is $1/\mu$. If we knew σ_d, then the proportion of time Channel d is busy would be $\xi_d = (1/\mu)\,/\sigma_d$. But ξ_d is already given by (20) and (17)! Hence, we reverse engineer σ_d as

$$\sigma_d = E[\text{Duration of Cycle } d] = \frac{1/\mu}{\xi_d} = \frac{1}{\lambda\,\phi_d} \tag{21}$$

So, the expected duration Channel d is free is $(\lambda\phi_d)^{-1} - \mu^{-1} = (\xi_d^{-1} - 1)/\mu$, which answers Question 3. In particular, the expected free time of Channel 1, is $1/\lambda$.

In light of the priority ordering of the channels, one anticipates that a channel with a higher ID will have longer free time (or equivalently, longer cycle time) on average than a channel with a lower ID; that is, $\sigma_1 < \sigma_2 < \ldots < \sigma_c$. To verify this inequality, in view of (21), it suffices to check that $\phi_1 > \phi_2 > \ldots > \phi_c$, which we left for the reader to do.

Finally, we should point out that oftentimes it is possible to renumber the channels/servers periodically so that they are used more equitably. Then the average busy time for all c channels is given by $E[N_c]/c = \rho(1 - \theta_c)/c$.

4 Application

Here we give two examples to illustrate how the Erlang loss formulas are used to determine the number of channels to install and/or operate.

Example 4.1 (Photocopy Machines) A photocopy shop anticipates arrival of customers at intervals of two minutes on average, and expects each customer to use a photocopy machine for 10 minutes on average. How many photocopy machines should the shop owner install? Assume that the inter-arrival times are IID exponential(0.5) and the service times are IID exponential(0.1).

As a crude approximation, the owner needs at least 5 machines, since the arrival rate $\lambda = 0.5$ is $\rho = 5$ times larger than the service rate $\mu = 0.1$. The calculations given in Table 1 help the shop owner make a more refined decision.

With only 5 machines installed, the shop will lose a whopping 28.5% of potential customers. In order not to lose any more than 10% of potential customers the owner should have 8 machines on operation. In that case, the shop will actually lose only about 7% of potential customers. But then the eighth machine will remain idle for almost 75% ($\approx 1 - 0.252$) of the time. Of course, in order to make equitable use of each machine the shop owner can cyclically renumber the machines each day. Then the average idle time of 8 machines will be about 42% ($\approx 1 - 0.581$), which will suffice to accommodate down time for machine maintenance.

Table 1 Quantities essential for decision making, when $\rho = 5$

Machine d	1	2	3	4	5	6	7	8	9
ω_d	0.833	0.676	0.530	0.398	**0.285**	0.192	0.121	**0.070**	0.037
ϕ_d	0.167	0.158	0.146	0.131	0.113	0.093	0.071	0.050	0.033
$E[N_d]$	0.833	1.622	2.352	3.008	3.576	4.041	4.397	4.650	4.813
ξ_d	0.833	0.788	0.730	0.657	0.567	0.465	0.357	**0.252**	0.163
$E[N_d]/d$	0.833	0.811	0.784	0.752	0.715	0.673	0.628	**0.581**	0.535

Source Computed using R codes

Table 2 Quantities essential for decision making, when $\rho = 15$

Cab d	13	14	15	16	17	18	19	20	21	22	23	24
ω_d	0.263	0.220	0.180	0.145	0.113	0.086	0.064	0.046	**0.032**	0.021	0.014	0.008
ξ_d	0.696	0.649	0.595	0.536	0.472	0.405	0.337	0.272	0.211	0.157	0.113	0.077
$E[N_d]/d$	0.850	0.836	0.820	0.802	0.783	0.762	0.739	0.716	**0.692**	0.668	0.643	0.620

Source Computed using R codes

Example 4.2 (Taxicab Drivers) A taxicab company owns a fleet of 24 cabs. The company wants to ensure that taxicabs will wait idle for approximately 30% of duty time *on average*. This will give sufficient rest for the driver and the cab, but not waste too much time idling. The company models a client's ride time as exponential with mean 30 minutes. If there is an exponential waiting time with mean two minutes between ride requests from clients, how many taxicab drivers should the company place on duty?

Here, $\rho = 30/2 = 15$. But with only 15 cab drivers on duty, the average idle time will be only 18%, and the company will lose about 18% of potential customers. See Table 2. The company should put 21 drivers on duty to ensure an average idle time of about 30.8%, and then the company will lose only 3.2% of potential customers.

5 Conclusion

We proved the Erlang loss formulas for the $M/M/c/c$ queuing system by using an intuitive limit theorem for an alternating renewal process and recursive relations, which we solved using mathematical induction. These formulas extend easily to the $M/M/d/d$ sub-system (consisting of Channels 1 through d) for $1 \leq d \leq c$. Thereafter, we have used these formulas, and other quantities derived from them, to answer several practically useful questions in order to make decisions. The simplicity of our approach will make the Erlang loss formmulas comprehensible to beginning college mathematics students. It is our humble tribute to Agner Krarup Erlang shortly after the centenary year of his celebrated formulas.

We should mention that the recursive relations we have used hold when both the inter-arrival times are exponential(λ) and the service times are exponential(μ). It is possible to extend our method to the case when the inter-arrival times are IID having an Erlang distribution (that is, a gamma distribution with an integer-valued shape parameter) and/or when the service times are IID having another Erlang distribution. In fact, using more advanced methods, it has been shown (see Takacs (1969)) that the Erlang loss formulas hold even when service times are IID with any arbitrary continuous distribution.

Acknowledgements Thanks to my students who participated in a discussion of this problem. Thanks to Professor Bikas K. Sinha for providing feedback on an earlier manuscript.

References

Abate, J., & Whitt, W. (1998). Calculating transient characteristics of the Erlang loss model by numerical transform inversion. *Communications in Statistics. Stochastic Models*, *14*(3), 663–680.

Mann, N. R., Schafer, R. E., & Singpurwalla, N. D. (1974). *Methods for statistical analysis of reliability and life data*. New York-London-Sydney: Wiley Series in Probability and Mathematical Statistics, John Wiley & Sons.

Medhi, J. (1982). *Stochastic processes*. New York: A Halsted Press Book. John Wiley & Sons.

Messerli, E. J. (1972). Proof of a convexity property of the Erlang B formula. *The Bell System Technical Journal*, *51*, 951–953.

O'Connor, J. J., & Robertson, E. F. (xxxx). *MacTutor history of mathematics archive*. http://www-history.mcs.st-andrews.ac.uk/Biographies/Erlang.html

Rausand, M., & Høyland, A. (2004). *System reliability theory. Models, statistical methods, and applications* (2nd ed.). Hoboken, NJ: Wiley Series in Probability and Statistics. Wiley-Interscience.

Ross, S. M. (1996). *Stochastic processes*, (2nd ed.). New York: Wiley Series in Probability and Statistics, John Wiley & Sons.

Sarkar, J., & Li, F. (2006). Limiting average availability of a system supported by several spares and several repair facilities. *Statistics & Probability Letters*, *76*(18), 1965–1974.

Sarkar, J. (2006). Random walk on a polygon, recent developments in nonparametric inference and probability, IMS lecture. *Notes*, *50*, 31–43.

Takacs, L. (1969). On Erlang's formula. *The Annals of Mathematical Statistics*, *40*, 71–78.

Machine Learning, Regression and Optimization

Biswa Nath Datta and Biswajit Sahoo

Abstract Machine learning is a subfield of artificial intelligence (AI). While AI is the ability of the machine to think like humans, machine learning is the ability of machine to learn from data without any explicit instructions. Applications of machine learning are abundant: stock-price forecast; face, speech and handwriting recognition; medical diagnosis of diseases like cancer, blood pressure, diabetes, neurological disorders including autism, spinal stenosis and others; and health monitoring, just to name a few. Potential applications of machine learning in solutions to many other complex practical problems are currently being investigated. An ultimate goal of machine learning is to make predictions based on a properly trained model. Two major techniques of supervised machine learning are: statistical regression and classification. For best prediction, the parameters of the model need to be optimized. This is an optimization task. After giving a brief introduction to machine learning and describing the role of regression and optimization, the paper discusses in some detail the basics of regression and optimization methods that are commonly used in machine learning. The paper is interdisciplinary, blending machine learning with statistical regression and numerical linear algebra, and optimization. Thus, it will be of interest to a wide variety of audiences ranging from mathematics, statistics and computer science to various branches of engineering.

Keywords Machine learning · Regression · Least-squares estimator · Optimization

The paper is based on a Keynote Talk delivered by Biswa Nath Datta at the International Conference on "Data-Science and SDGS: Challenges, Opportunities and Realities", held in University of Rajshahi, Bangaldesh, and subsequent Plenary Talks at IIT-Kharagpur, India, and Sri Padmavati Mahila Visvavidyalayam, Tirupati, and a Colloquium Talk at VJTI, Mumbai, India.

B. N. Datta (✉)
Department of Mathematical Sciences, Northern Illinois University, De Kalb, IL 60115, USA

Department of Mathematics, Indian Institute of Technology, Kharagpur, India

B. Sahoo
Department of Mechanical Engineering, Indian Institute of Technology, Kharagpur 721302, West Bengal, India
e-mail: biswajitsahoo1111@iitkgp.ac.in

B. K. Sinha et al. (eds.), *Data Science and SDGs*,
https://doi.org/10.1007/978-981-16-1919-9_15

1 Introduction

Machine learning is defined as the scientific study of algorithms and statistical models that computer systems use to perform a specific task without using explicit instructions, relying on patterns and inference instead (Wikipedia). It is a subset of artificial intelligence (AI). The goal of AI is to build systems that are capable of independent reasoning without human supervision. Though machines are yet to achieve artificial general intelligence, they have become particularly good at a subset of specialized tasks. The excellent performance of machines at certain tasks is achieved by using machine learning algorithms. An exponential increase in computing power and the availability of abundant data have led to the phenomenal success of machine learning. Increasingly, machine learning is being applied to yet unexplored areas of applications.

Learning of machines is achieved in several sequential steps. An important step is "*Model Building*". This task involves building an appropriate model from the given set of data. This model is then used to make predictions, which is the ultimate goal of machine learning. Two major approaches of model building used in supervised machine learning (see later for a description of this learning) are *Regression* and *Classification*. Regression is a well-known statistical technique used to build a mathematical model relating to the input and output data. In machine learning, regression is used when the outcome is continuous. Model building for the classification is similar to regression—it is used to categorically separate the data into two or more classes. Classification is thus used in machine learning to predict discrete outcomes.

In order to make a prediction as good as possible, the parameters of the learned model must be fine-tuned. This is an optimization task. The task involves minimizing or maximizing a well-defined objective function with the parameters to be optimized as unknown variables. For example, if the regression model is a straight line, then the two parameters, slope and intercept, need to be chosen in such a way that an objective function (called cost) involving these two parameters as variables is minimized. Similar remarks hold for other regression and classification models.

Regression and optimization are two well-established areas of study and research, and both arise in a wide variety of applications in science, economics, engineering and industry. There are well-known books on both these topics, such as Montgomery et al. (2003), Nocedal and Wright (2006). This paper gives a brief introduction to basic regression and optimization methods that are routinely used in machine learning. A special emphasis is given to numerical linear algebra concepts and numerically viable algorithms, which are absolutely essential to the understanding of the material of this paper. These include the *QR* and *SVD methods* for *solving the normal equations*, and the optimization techniques, such as the *Gauss–Newton* and *Levenberg–Marquardt* methods for *nonlinear regression*. These concepts and techniques may not be readily available to the general machine learning community. Regrettably, because of the limited scope of this paper, and space limitations, several important topics of machine learning, such as, *Cross Validation* and the *Stochastic Gradient methods* for *Regression*, and *Logistic Regression* and *Support Vector Machine* techniques for classification cannot be included. These topics and many others can be easily found in any textbook on machine learning, see Watt et al. (2020), and Bishop (2006), besides, the internet is also a good source. Plenty of information is available on the internet. For details of various aspects of machine learning and deep learning, the interested readers are referred to Bishop (2006), Goodfellow et al. (2016), Creswell et al. (2018), Krizhevsky et al. (2012), and Raschka et al. (2020).

1.1 Programming Versus Machine Learning

A computer program executes a set of explicit instructions (rules) to perform a particular task. Therefore, to solve a problem, one has to meticulously specify all the rules that the computer needs to execute to arrive at a conclusion. While this may be an effective strategy for simple problems, it turns out that for complex problems, it performs poorly. This is for the reason that in complex scenarios, it is not possible to hand-code all the rules, as some of the rules may not be known beforehand.

In contrast, in machine learning, the computer learns the underlying rules from data. If enough data are available, using machine learning techniques, it is possible to learn the rules no matter how complex they are. This has enabled computers to produce excellent results at certain complex tasks that are nearly impossible to solve using traditional programming. Some of the important application areas of machine learning are discussed in Sect. 1.5 (Step 4).

1.2 Data Science and Machine Learning

Data are central to both data science and machine learning. The performance of machine learning system is dependent on the quality of the data used to train the system. In practice, the data are usually unstructured and contain missing entries as well as outliers. Therefore, the data should be cleaned and preprocessed to a format that can then be used in machine learning. Data science is more inclined toward preparing the data and gaining insights from it. Whereas, machine learning is concerned with the development of models and algorithms to extract useful information from the data. However, there is considerable overlap between the two terms and most of the time, both the terms are used interchangeably.

1.3 Some Essential Numerical Optimization Concepts and Techniques for Machine Learning

As stated before, machine learning requires parameters of the model to be optimized for best performance. This is an optimization task. It involves optimizing (minimizing or maximizing) certain cost function involving the parameters of the model. Optimization problems are classified as: *Unconstrained* and *Constrained*. Both of these are of interest to two major approaches used for supervised model training; *regression* and *classification.*

Let $F(\boldsymbol{X})$, where $\boldsymbol{X} \in \mathbb{R}^n$ represents the function of the training model ($F : \mathbb{R}^n \rightarrow \mathbb{R}$). Then, the unconstrained minimization problem is stated as

$$\underset{x}{\boldsymbol{minimize}}\, F(\boldsymbol{X})$$

Before stating the optimality conditions for unconstrained optimization, we first define the following concepts:

- Gradient of $F(\boldsymbol{X}) = \nabla F(\boldsymbol{X}) = \left[\frac{\partial F}{\partial x_1}, \frac{\partial F}{\partial x_2}, \ldots, \frac{\partial F}{\partial x_n} \right]^T$ (A vector)
- Hessian of $F(\boldsymbol{X}) = \nabla^2 F(\boldsymbol{X}) = \boldsymbol{H}_F(\boldsymbol{X}) = \left(\frac{\partial^2 F}{\partial x_i \partial x_j} \right)_{n \times n}$ (A matrix)

where, $\frac{\partial F}{\partial x_i}$ = first derivative of F with respect to x_i, $\frac{\partial^2 F}{\partial x_i^2}$ = second-order partial derivative of F with respect to x_i, and $\frac{\partial^2 F}{\partial x_i \partial x_j} = \frac{\partial F}{\partial x_i}\left(\frac{\partial F}{\partial x_j} \right)$.

- If F maps to a higher dimensional space, i.e., ($F : \mathbb{R}^n \rightarrow \mathbb{R}^m$), the Jacobian matrix $J(\boldsymbol{X}) = \left(\frac{\partial F}{\partial x_i} \right)_{m \times n}$ (A matrix).
- The Hessian matrix is a matrix of second-order partial derivatives and the Jacobian is a matrix of the first-order partial derivatives of $F(\boldsymbol{X})$.

Optimality Conditions for Unconstrained Minimization

Let $\boldsymbol{X}^*$ be a critical point of $F(\boldsymbol{X})$; that is $\nabla F(\boldsymbol{X}^*) = 0$. Then,

- A necessary condition for $\boldsymbol{X}^*$ to be a local minimizer is that the Hessian matrix $\boldsymbol{H}_F(\boldsymbol{X}^*)$ is positive semidefinite. This means that *for a critical point* $\boldsymbol{X}^*$ *to be local minimizer,* $\boldsymbol{H}_F(\boldsymbol{X}^*)$ *must be positive semidefinite.*
- A sufficient condition for $\boldsymbol{X}^*$ to be local minimizer is that of the Hessian matrix $\boldsymbol{H}_F(\boldsymbol{X}^*)$ is positive definite. This means that if $\boldsymbol{X}^*$ *is a critical point of* $F(\boldsymbol{X})$, *and if* $\boldsymbol{H}_F(\boldsymbol{X}^*)$ *is positive definite, then it is a local minimizer* (In fact a *strict local minimizer*).

Convexity and Optimization

- A function $F(\boldsymbol{X})$ is **convex** if its domain is a convex set and for all $\boldsymbol{X}_1$ and $\boldsymbol{X}_2$ in the domain of $F(\boldsymbol{X})$, and $\alpha \in [0, 1]$, the following inequality holds:

$$F(\alpha \boldsymbol{X}_1 + (1-\alpha)\boldsymbol{X}_2) \leq \alpha F(\boldsymbol{X}_1) + (1-\alpha)F(\boldsymbol{X}_2)$$

- If the inequality is strict, then F is strictly convex.
- An important property of convex optimization: *A local minimizer of a convex optimization problem is also a global minimizer.*

Solving a Convex Optimization Problem

For a convex problem, the necessary conditions of optimality are also sufficient. Thus, a local (therefore, global) minimizer of an unconstrained convex function $F(\boldsymbol{X})$ can be found simply by solving the system of equations given by $\nabla F(\boldsymbol{X}) = 0$. This makes solving such problems a lot easier.

Characterization of a Convex Function

- $F(\boldsymbol{X})$ is convex if and only if the Hessian matrix $\boldsymbol{H}_F(\boldsymbol{X})$ is positive semidefinite for all $\boldsymbol{X}$.
- If $\boldsymbol{H}_F(\boldsymbol{X})$ is positive definite, $F(\boldsymbol{X})$ is strictly convex and conversely.

(A symmetric matrix is positive semidefinite (positive definite) if and only if all the eigenvalues are nonnegative (positive)).

Constrained Convex Optimization

A constrained minimization problem of the form

$$\begin{aligned}
&\textit{minimize } F(\boldsymbol{X}) \\
&\textit{subject to, } G_i(\boldsymbol{X}) \leq 0, \ i = 1, 2, \ldots, m \\
&\textit{and } H_j(\boldsymbol{X}) = 0, \ j = 1, 2, \ldots, n
\end{aligned}$$

is convex if (i) the functions $F(\boldsymbol{X})$ and $G_i(\boldsymbol{X})$ are convex and (ii) $H_j(\boldsymbol{X})$ are affine. The **portfolio optimization problem** considered in this paper is a constrained convex problem (see later for the statement of this problem).

Solution of the Constrained Convex Optimization Problem

A constrained convex optimization problem, like the unconstrained one, can also be solved just from the necessary conditions of optimality for this problem, popularly known as the *Karush-Khun-Tucker* (KKT) conditions. For page restrictions, we omit the details of the KKT conditions, see Nocedal and Wright (2006).

Numerical methods for Solving Optimization Problems

The nonconvex unconstrained optimization problems, such as those encountered in the nonlinear regression problems, need to be solved using numerical algorithms; such as the *Secant Method*, *Newton's method*, the *quasi-Newton methods* (e.g., the BFGS algorithm), etc.

- Starting with an initial approximation $\boldsymbol{X}_0$, these iterative methods construct a sequence of solutions $\{\boldsymbol{X}_k\}$, defined by $\boldsymbol{X}_{k+1} = \boldsymbol{X}_k + \boldsymbol{S}_k,\ k = 0, 1, 2, \ldots$ where $\boldsymbol{S}_k$ is called the *direction vector*, with a hope that these iterations will converge to a solution under certain conditions. Different methods differ in the way the direction vectors $\boldsymbol{S}_k$'s are computed. A common and an obvious criterion for termination of the iteration is: Stop if $\|\nabla F(\mathbf{X}_k)\| \leq \epsilon$, for a pre-chosen small enough ϵ. Below, we just state how the direction vector for *Newton's method is chosen*, because the methods for *nonlinear regression* (see later) are derived from this method.

Choosing the Direction Vector for Newton's Methods

For the popular Newton's method, which has a *quadratic* rate of convergence (If $\boldsymbol{X}_0$ is chosen sufficiently close to the solution), the vector $\boldsymbol{S}_k$ is computed by solving the system of equations:

$$\boldsymbol{H}_F(\boldsymbol{X}_k)\boldsymbol{S}_k = -\nabla \mathrm{F}(\mathbf{X}_\mathrm{k})$$

where $\nabla \mathrm{F}(\mathbf{X}_\mathrm{k})$ and $\boldsymbol{H}_F(\boldsymbol{X}_k)$ are the values of the gradient vector and the Hessian matrix of $F(\boldsymbol{X})$ respectively, at $\boldsymbol{X} = \boldsymbol{X}_k$. This system is linear if $F(\boldsymbol{X})$ is linear. The details can be found in Nocedal and Wright (2006), or in the upcoming book by Datta (2020). MATLAB function *fsolve* can be used to solve the above system of equations. Newton's method *works poorly* if $\boldsymbol{H}_F(\boldsymbol{X})$ is *ill-conditioned*; in that case, a variation of the method such as the *BFGS method* needs to be used.

1.4 Types of Learning

Learning can broadly be divided into two categories:

1. **Supervised learning**: In supervised learning, both input and output data are fed into the system with the hope that after seeing enough examples, the machine would be able to generate correct outputs for previously unseen inputs.
2. **Unsupervised learning**: In unsupervised learning, a computer is fed only the input data. The computer is expected to find hidden patterns in the data itself and categorize the input data according to hidden patterns. An important example of unsupervised learning is *clustering*.

Besides supervised and unsupervised learning, there is another class of learning called *reinforcement learning*. In reinforcement learning, instead of the output, the system is given a reward or a penalty at each step as feedback as it performs its task. In technical terms, the system (also known as agent) tries to achieve its goal by adopting a policy that maximizes the reward. This type of learning is mainly used in **computer games** and **autonomous systems**. *We will discuss only supervised learning with a special attention to regression techniques.*

1.5 Basic Steps of Machine Learning

The learning and testing processes of machine learning are achieved by performing several sequential steps. These steps are as follows. A diagrammatic representation of these steps is given in Fig. 1.

Step 1: Problem Definition

Every machine learning problem needs to be defined clearly and unambiguously.

Step 2: Data Collection

As the machine learns the rules from the data, relevant data need to be collected for the task at hand. The data are divided into two sets: Training Set and Test Set. Machine learning model is trained on the training data and its performance is evaluated on the test data. The goal of every machine learning model is to perform well on the previously unseen data. The Test Set acts as a new set as if that has not been previously seen by the machine during its training process. If the model performs well on the test data, it is expected to perform well on other new data as well. Finally, if the data are collected for a supervised task, both the input and output data are collected. For unsupervised task, only the input data are collected.

Step 3:Feature Design

This step consists of extracting the important features from the raw data, which are most relevant to the problem to be solved. The choice of the right features facilitates the learning process and is thus a very important step. For example, if the problem is to distinguish a cat from a dog, then the features, like the animal has four legs or two ears or two eyes, do not obviously help the learning process. However, the features like size of the nose and shape of the ears are quite relevant and are important features for this problem. Similarly, if the problem is forecasting the stock price of a

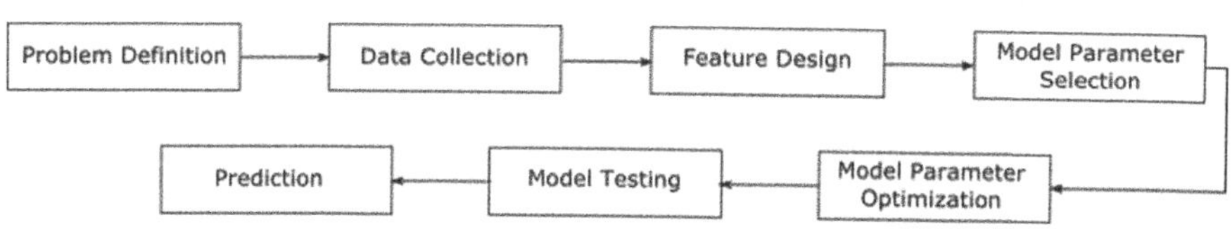

Fig. 1 Steps of machine learning. *Source* Created by the authors

company, the most important feature is the revenue of the company, among others. Research is currently being undertaken to make this process automated.

Note: It is noted that for certain types of machine learning techniques such as **Deep Learning** see Goodfellow et al. (2016), this step is not required. Instead, important hidden features are learned by the algorithm from the data itself.

Step 4:Selecting Appropriate model and Parameters

This step involves training the machine to find an appropriate mathematical model that maps the inputs of the training set to the outputs. The two common techniques for supervised learning are: **Regression** and **Classification**.

Regression is used to predict the outcome when the output is continuous. The most common everyday life examples of regression include: *Predicting the Stock Price forecast* of a company from its past revenues; *Sales Price Analysis* of a certain product from the data on past sales or from changes of certain variable affecting the sales (e.g., GDP); *Diagnosis of certain diseases*, such as blood pressure, heart disease from the respective quantitative measurements of blood glucose and cholesterol levels, etc., over a certain period of time and many others.

Classification is used when the output is discrete. As the name suggests, this is a technique to classify the training data into two or more categories according to designed features. The most common examples include: *Object Detection in Computer Vision and Image Processing*; *Sentiment Analysis*: Analysis of public sentiment expressed on social media toward a recently released product (e.g., a movie) to determine the possibility of success or failure of the product; *Medical Diagnostic Tool*: Detection of certain neurological disorders, such as Autism, Attention Deficit Hyperactivity Disorder (ADHD), Spinal Stenosis. This technique is also used to detect if a patient is predisposed to a certain genetic cancer (e.g., Breast Cancer); *Spam Filtering*: Classifying an e-mail or a text message as spam; *Digital Hand-Writing Recognition*: Routinely used in mobile banking.

Step 5:Model Parameter Optimization

For best performance, the model parameters should be optimized using some appropriate optimization techniques. The optimization of the parameters of a "Regression Model" leads to the least-squares minimization (linear and nonlinear) of the objective (cost) function. *The linear least-squares problems are convex and can be solved just by solving a linear system of equations.* For nonlinear least-squares problems, some optimization techniques need to be used (for details, see Sect. 4). Similarly, the optimization of the parameters of the line or a hyperplane of the "Classification Model" that separates the test data into two or more classes leads to optimization problems, which are to be solved using numerical optimization algorithms, see Watt et al. (2020).

Step 6:Model Testing

Once the model is trained, the final step is to test the efficacy of the model. If the results are not acceptable, then the model needs to be adjusted by using different design features and/or including more training data. The test data set which was set aside or a new set of data with known results can be used for testing purposes. For linear regression models, mean-squared error (MSE) is a common testing criterion.

Step 7:Model Prediction

The whole purpose of machine learning is to make a prediction based on the learned model. Prediction is done on new data given to the computer.

2 Linear Regression Analysis and Least-Squares Techniques

"Regression analysis" is a well-known statistical technique used to model or establish a relationship between variables of a set of input–output data.

2.1 *Multivariable Linear Regression*

Let $\boldsymbol{X}_1, \boldsymbol{X}_2, \ldots, \boldsymbol{X}_k$ be a set of k vector-valued inputs assumed to be related to the output $\boldsymbol{Y}$. Then the multivariable linear regression problem is to find the *regression parameters*, $\beta_0, \beta_1, \ldots, \beta_k$, such that

$$y_i = \beta_0 + \beta_1 x_{i1} + \beta_2 x_{i2} + \cdots + \beta_k x_{ik} + \epsilon_i, \ i = 1, 2, \ldots, n$$

where, $\boldsymbol{X}_j = \left[x_{1j}, x_{2j}, \ldots, x_{nj}\right]^T, \ j = 1, 2, \ldots, k$ ***and*** $\boldsymbol{Y} = [y_1, y_2, \ldots, y_n]^T$. Using the notation 1 to represent a vector of all ones of size $(n \times 1)$, the above equation, in matrix–vector form, can be written as: $\boldsymbol{Y} = \boldsymbol{X\beta} + \boldsymbol{\epsilon}$, where $\boldsymbol{X} = [1, \boldsymbol{X}_1, \boldsymbol{X}_2, \ldots, \boldsymbol{X}_k]$ is a matrix of size $(n \times (k+1))$, $\epsilon = [\epsilon_1, \epsilon_2, \ldots, \epsilon_n]^T$, ***and*** $\boldsymbol{\beta} = [\beta_0, \beta_1, \beta_2, \ldots, \beta_k]^T$.

Statistical Assumptions: (i) $E[\boldsymbol{\epsilon}] = \boldsymbol{0}$, and (ii) $Cov(\boldsymbol{\epsilon}) = \sigma^2 \boldsymbol{I}$. These assumptions are needed to guarantee some desirable properties of the estimator $\widehat{\boldsymbol{\beta}}$ (see Sect. 2.4).

2.2 *Least Squares Solution of the Regression Model*

Step 4 of the machine learning problem requires that the regression parameters $\beta_0, \beta_1, \ldots, \beta_k$ are to be chosen in such a way that the model is the best fit. *It has been established in* Montgomery et al. (2003)*that the regression parameters obtained as the solution of the following least-squares problem (LSP) represent the best fit to the data* (see the properties of the least-squares estimation listed below).

Define the least-squares function:

$$S(\boldsymbol{\beta}) = S(\beta_0, \beta_1, \ldots, \beta_k) = \sum_{i=1}^{n}\left(y_i - \beta_0 - \sum_{j=1}^{k}\beta_j x_{ij}\right)^2 = \sum_{i=1}^{n}\epsilon_i^2.$$

Then the least-squares problem is to find the vector $\boldsymbol{\beta}$ such that $S(\boldsymbol{\beta})$ is minimized:

$$Minimize\ S(\boldsymbol{\beta}) = \|\boldsymbol{\epsilon}\|_2^2 = \|\boldsymbol{X\beta} - \boldsymbol{Y}\|_2^2.$$

Thus, finding a least-squares estimator of $\boldsymbol{\beta}$, denoted by $\widehat{\boldsymbol{\beta}}$, *amounts to minimizing the square of the 2-norm of the error vector* $\boldsymbol{\epsilon}$*(that is why the problem is so-called).*

2.3 Solution of the Least-Squares Problem in Optimization Setting

The above is an unconstrained minimization problem. Standard numerical methods, such as the *steepest descent, Newton and Quasi-Newton,* or the *conjugate gradient* (for large problems), etc., can be used to minimize the function $S(\boldsymbol{\beta})$. However, it turns out that the least-squares function $S(\boldsymbol{\beta})$ is *convex*; this is because, the Hessian matrix, $H_S(\boldsymbol{\beta}) = 2\boldsymbol{X}^T\boldsymbol{X} \succcurlyeq 0$, is *symmetric positive semi-definite.* According to a well-known characterization of convex function (see Sect. 1.3), $H_S(\boldsymbol{\beta})$ is a convex function. Thus, the function $S(\boldsymbol{\beta})$ can be *minimized just by solving the system of equations* given by $\nabla S(\boldsymbol{\beta}) = 2\left(\boldsymbol{X}^T\boldsymbol{X\beta} - \boldsymbol{X}^T\boldsymbol{Y}\right) = 0$. That is, the least-squares solution $\boldsymbol{\beta}$ satisfies the so-called *Normal Equations*: $\boldsymbol{X}^T\boldsymbol{X\beta} = \boldsymbol{X}^T\boldsymbol{Y}$, see Datta (2010). Furthermore, if $\boldsymbol{X}$ has full rank, then $\boldsymbol{X}^T\boldsymbol{X}$ is symmetric positive definite, and therefore, is nonsingular. So, $\boldsymbol{\beta}$ is unique in this case. The analytic solution for the *full rank normal equations* is: $\boldsymbol{\beta} = \left(\boldsymbol{X}^T\boldsymbol{X}\right)^{-1}\boldsymbol{X}^T\boldsymbol{Y}$.

Pseudo-inverse and the least-squares estimator:

The matrix $\left(\boldsymbol{X}^T\boldsymbol{X}\right)^{-1}\boldsymbol{X}^T$ is called the *pseudo-inverse* of $\boldsymbol{X}$, denoted by $\boldsymbol{X}^\dagger$, thus, $\widehat{\boldsymbol{\beta}} = \boldsymbol{X}^\dagger\boldsymbol{Y}$ (Rao and Mitra 1971). However, $\widehat{\boldsymbol{\beta}}$ should never be computed by inverting the matrix $\boldsymbol{X}^T\boldsymbol{X}$ explicitly (see Datta 2016) for the purpose of least-squares estimation. Having said this, it is noted that its explicit computation might be required for other statistical computations, see Montgomery et al. (2003), Freedman (2009), and Rencher and Schaalje (2008).

2.4 *Properties of the Least-Squares Estimator*

It can be shown that under the statistical assumptions stated in Sect. 2.1, the least-squares estimate enjoys the following important properties: $E\left[\widehat{\boldsymbol{\beta}}\right] = \boldsymbol{\beta}$, and $Cov\left(\widehat{\boldsymbol{\beta}}\right) = \sigma^2\left(X^T X\right)^{-1}$. That is, the **least-squares estimator $\widehat{\boldsymbol{\beta}}$is the best linear unbiased estimator** in the sense that it has the minimum variance among all other linear unbiased estimators (**Gauss–Markov theorem**) (Freedman 2009).

3 Computational Algorithms for Computing Linear Least-Squares Solution

In this section, we state three computationally effective algorithms to solve the problem.

3.1 *The Cholesky Method for Least Squares Solution*

This method is based on the Cholesky factorization of the matrix $\boldsymbol{X}^T\boldsymbol{X}$ and should be used only when the matrix $\boldsymbol{X}$ is well conditioned.

The Cholesky Algorithm

Step 1. Find the Cholesky factorization of the symmetric positive definite matrix: $\boldsymbol{X}^T\boldsymbol{X} = \boldsymbol{R}^T\boldsymbol{R}$; where $\boldsymbol{R}$ is an upper triangular matrix, called the *Cholesky factor*. MATLAB function *chol* computes the Cholesky factor.

Step 2. Solve the lower triangular system: $\boldsymbol{R}^T\boldsymbol{p} = z$, where $z = \boldsymbol{X}^T\boldsymbol{Y}$.

Step 3. Solve the upper triangular system to compute the least-squares estimate: $\boldsymbol{R}\widehat{\boldsymbol{\beta}} = \boldsymbol{p}$.

Ill-Conditioning of *X*and the QR and the SVD Methods

If the matrix $\boldsymbol{X}$ is ill-conditioned, that is, if the smallest singular value of $\boldsymbol{X}$, $\sigma_{min}(\boldsymbol{X})$, is too small, then $\boldsymbol{X}^T\boldsymbol{X}$ will be severely ill-conditioned, since $Cond\left(\boldsymbol{X}^T\boldsymbol{X}\right) = [Cond(\boldsymbol{X})]^2$. In this case, the Cholesky algorithm will not yield an accurate result, the $\boldsymbol{QR}$ and SVD methods, stated below should be used, instead. The SVD method should be used if the matrix X is highly ill-condition, otherwise, the QR method is good enough for all practical purposes, see Datta (2010).

3.2 *The QR Factorization Method for Least-Squares Solution*

The QRFactorization Algorithm

Idea: Transform the least-squares problem, by means of an orthogonal transformation, to a problem of solving an upper triangular system based on the QR factorization of the matrix X.

Step 1. Factorize X into QR: $X = QR$, where $Q \in \mathbb{R}^{n\times n}$ and $R \in \mathbb{R}^{n\times k}$; partition Q and R as shown below.

$$\text{MATLAB function: } \left[Q, R\right] = qr(X), \text{ where } Q = \left[Q_1, Q_2\right], \quad R = \begin{bmatrix} R_1 \\ 0 \end{bmatrix},$$

where R_1 is an $(k \times k)$ nonsingular upper triangular matrix, and Q_1 is an $(n \times k)$ matrix with orthonormal columns.

Step 2. Solve the $(k \times k)$ *upper triangular* system to obtain the least-squares estimate $\widehat{\beta}$: $R_1\widehat{\beta} = z$, where $z = Q_1^T Y$.

Note that $X^T X\beta = X^T Y \Rightarrow R_1\widehat{\beta} = Q_1^T Y = z$ (since Q is orthogonal).

3.3 *The SVD Method for Least-Squares Solution*

The SVD Algorithm

Idea: Transform the problem to a problem of solving a diagonal system, using the SVD (Singular Value Decomposition) of X.

Step 1. Compute the SVD of X: $X = USV^T$, where U and V are *orthogonal matrices* of order n and k respectively, and S is a rectangular *diagonal* matrix of order $(n \times k)$. MATLAB function: $[U, S, V] = svd(X)$, where S is a diagonal matrix and U, V are orthogonal matrices.

Step 2. Solve the diagonal system $Sz = b$, where $b = U^T Y$.

Step 4. Obtain the least-squares estimate $\widehat{\beta} = Vz$.

Note: As in the QR method, the skinny form of U and S can be used for the SVD method as well. This would save computational time, but will yield the same result, see Datta (2010).

4 Nonlinear Regression Models

A general nonlinear model can be represented as

$$Y = F(X, \Theta) + \epsilon$$

where, F is a *nonlinear function* and Θ is a vector of unknown parameters of size $(k \times 1)$. The nonlinear least-squares problem for this model is:

$$\min_{\Theta} S(\Theta) = \frac{1}{2} R(\Theta)^T R(\Theta)$$

where, $R(\Theta) = [R_1(\Theta), R_2(\Theta), \ldots, R_n(\Theta)]$ ***and*** $R_i(\Theta) = Y_i - F(X_i, \Theta)$, $i = 1, 2, \ldots, n$. Unfortunately, this *function is not convex.* Thus, a numerical method needs to be used. **Newton's method** is widely used for its quadratic rate of convergence (provided that the initial approximation is close to the actual solution). However, each iteration of this method requires computation and solution of a system involving a Hessian matrix and thus, the method becomes computationally prohibitive.

The following alternatives to Newton's method aim at transforming the nonlinear least-squares problem to a linear least-squares problem involving only with the Jacobian of $S(\boldsymbol{\Theta})$, *which make them computationally attractive and practical.*

4.1 The Gauss–Newton Method

Starting from an initial approximation Θ_0, the k-th Newton iteration computes the *direction vector* S_k and it is then used to update the next approximation:

$$\Theta_{k+1} = \Theta_k + S_k$$

While Newton's method computes this vector by solving a system with a Hessian matrix, the Gauss–Newton method computes S_k by solving the linear least-squares problem

$$\underset{\Theta_k}{\boldsymbol{minimize}} \| J(\Theta_k) s_k + R(\Theta_k) \|_2^2$$

where, $J(\Theta_k)$ is the Jacobian of $S(\Theta)$ given by: $J(\Theta) = \left(\frac{\partial \Theta}{\partial \theta_i}\right)$ *(a matrix composed of the first-order partial derivatives of* Θ). The least-squares problem can now be solved using the *QR* or the SVD method, described in the preceding section.

4.2 The Levenberg–Marquardt Method

In this method, the Newton-direction vector S_k is computed by solving the *linear least-squares* problem:

$$\underset{\Theta}{minimize} \left\| \begin{bmatrix} J(\Theta_k) \\ \sqrt{\mu_k} I \end{bmatrix} s_k + \begin{bmatrix} R(\Theta_k) \\ \mathbf{0} \end{bmatrix} \right\|_2$$

where, μ_k is a suitably chosen parameter to overcome the difficulty of possible ill-conditioning of the Gauss–Newton least-squares problem. Again the QR or the SVD method can be used to solve this linear least-squares problem.

4.3 Results of Comparison of the Three Methods on a Toy Example (Heath 2018)

Given the following data (Table 1):

The problem is to find parameters θ_1 and θ_2, such that the curve $F(x, \Theta) = \theta_1 + e^{\theta_2 x}$ is the best fit to the data. The results of applying Newton's method, Gauss–Newton method and Levenberg–Marquardt method on this example are shown in the following figure (Fig. 2).

Observation: All the three methods produce the same result, accurate up to six decimal figures, see GitHub (2020) for Python implementation. Therefore, for all practical purposes, the solutions are identical. While Newton's method requires more computations for fine-tuning of the parameters, the *Gauss–Newton or the Levenberg–Marquardt achieves the same answer with less computations.*

5 Numerical Experiments

In this section, we present the results of our numerical experiments on Machine Learning with the following three real-life problems. The experiments were done using the popular programming language, Python. Python codes for these problems can be found in GitHub (2020). The data for these problems are available in the public domain.

Table 1 x- and y-coordinates of four data points

x	0	1	2	3
y	2	0.7	0.3	0.1

Source Heath (2018)

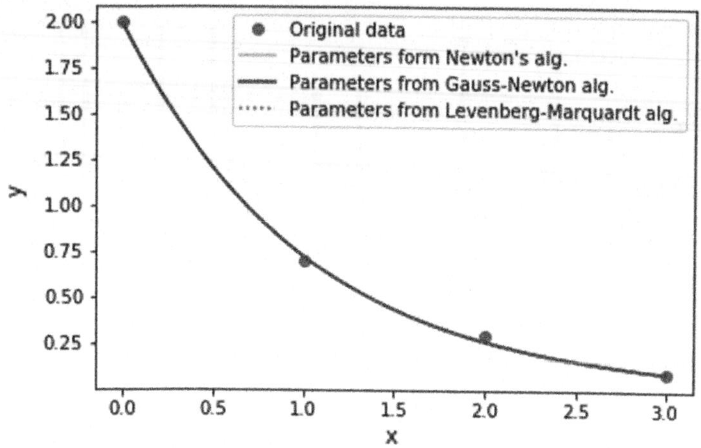

Fig. 2 Comparison of nonlinear fit using different methods. *Source* Created by the authors

5.1 Predicting House Price in the City of Boston

Problem Definition

The city of Boston in the USA is divided into different suburbs or towns. For simplicity, we will call them regions. The data on the median house prices are available in the public domain for a number of regions of this city along with other features about these regions. The problem is to predict median house prices of some new regions for which the data have not yet been collected.

Methodology

We divide the data into two sets: the training and the test set by randomly assigning 75% of the data to the training set and the rest to the test set. Then a multivariate linear regression model is fit to the training data and this model is then used to make a prediction on the test data. Results of the prediction are given in the following figure (Fig. 3). The x-axis of the figure goes from 1 to 126 as there are 126 regions in the test set.

Efficacy of the Model

The efficacy of the learned model is very often measured by using root mean squared error (RMSE) as the metric. RMSE for this problem is found to be 4.93. An RMSE value of zero would mean that all the predictions exactly match with the actual values in the test set. The aim of a prediction algorithm should be to reduce the RMSE value as much as possible. We achieved the present value of RMSE using multiple linear regression. Multiple linear regression fits a hyperplane to given training data points. Predictions are made by returning the value on the hyperplane corresponding to the test input variables. The model does not take the physical nature of the problem into consideration. Therefore, if predictions are made away from the given range of input

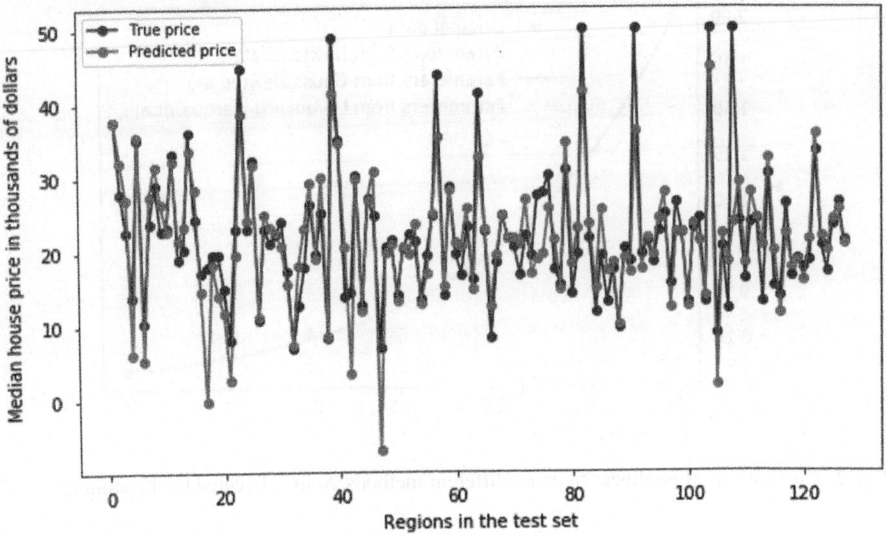

Fig. 3 Boston house price prediction. *Source* Created by the authors

variables, it might produce impractical results. It is possible to reduce RMSE even further by using a more complex model or a different machine learning technique.

5.2 *Predicting Number of Coronavirus (COVID-19) Cases in the State of New York*

Problem Definition

The problem here is to predict the number of cases of the Coronavirus in the state of New York in USA, one of the earlier hotspots in the country, from the past data.

Methodology

The first coronavirus case was detected in New York on March 1, 2020. Cumulative cases for successive dates were recorded. The data are publicly available. We use the data made available by New York Times (https://github.com/nytimes/covid-19-data (Accessed on 31 July, 2010)) from March 1, 2020 to June 14, 2020 for our study. The data for the first 90 days are taken as the training set and the remaining data as the test set. Both linear regression and piecewise linear regression techniques are applied to the training set and these trained models are then used to predict the number of future cases. The results are compared with the actual results from the test set.

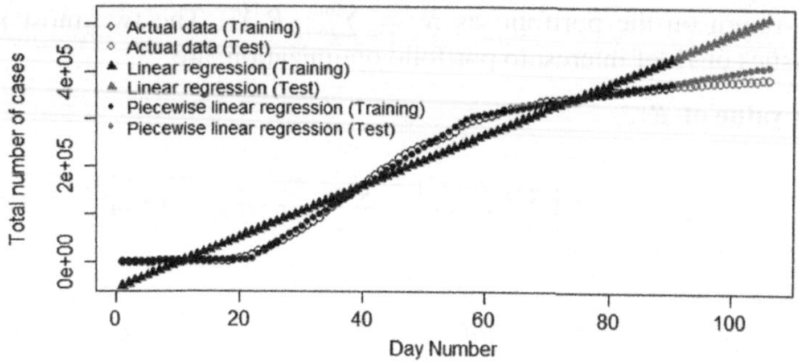

Fig. 4 Predicting the number of corona cases in New York. *Source* Created by the authors

Results

Predictions by both the linear regression model and piecewise linear regression model are plotted against the true values in the following figure (Fig. 4). We see that piecewise linear regression fit is more suitable for prediction, for this example, as compared with simple linear regression. This example also outlines one of the limitations of linear regression. *Observe that the values predicted by linear regression at the beginning are negative, which is impractical.*

5.3 *Portfolio Optimization*

This problem is concerned with optimizing certain desirable aspects of asset investments, such as maximizing the expected value of the return, minimizing the portfolio risk variance or both.

All these three problems are convex constrained quadratic optimization problems, and thus, can be solved just from the first-order necessity conditions (KKT conditions), which are computationally convenient and more efficient. Furthermore, each local solution is also a global solution in this case, because of the convexity (See Sect. 1.3).

Problem Definition

Suppose that there are n assets, $i = 1, 2, \ldots, n$ with respective rate of return R_i. Let X_i be the portion of the total investment to be invested in the asset i. Thus, if all the available funds are invested as well as there is no short selling, then

$$\sum_{i=1}^{n} X_i = 1; \ \ 0 \le X_i \le 1.$$

Define return on the portfolio as $R = \sum_{i=1}^{n} R_i X_i$. The two most important characteristics of R, of interest to portfolio optimization, are:

Expected value of *R*:

$$E[R] = E\left[\sum_{i=1}^{n} R_i X_i\right] = \sum_{i=1}^{n} X_i E[R_i] = \boldsymbol{X}^T \boldsymbol{m}$$

where, $\boldsymbol{X} = [X_1, X_2, \ldots, X_n]^T$ ***and*** $\boldsymbol{m} = [m_1, m_2, \ldots, m_n]^T$ ***and*** $m_i = E[X_i]$.

Variance of *R*:

$$Var(R) = E\left[(R - E[R])^2\right] = \sum_{i=1}^{n} \sum_{j=1}^{n} X_i X_j \sigma_i \sigma_j C_{ij} = \boldsymbol{X}^T \boldsymbol{C} \boldsymbol{X}$$

where, $\sigma_i = E[R_i - m_i]$, and $C_{ij} = E[(R_i - m_i)(R_j - m_j)]$.

Since the goal of an investor is to (i) minimize the total variance of R, or (ii) maximize the total expected value of R, or (iii) both, the following three optimization problems can be formulated.

Problem 1 (Risk (Variance) Minimization):

$$\begin{aligned} & minimize \frac{1}{2} \boldsymbol{X}^T \boldsymbol{C} \boldsymbol{X} \\ & subject\ to, \\ & \sum_{i=1}^{n} X_i = 1 \\ & \boldsymbol{X}^T \boldsymbol{m} = \boldsymbol{m}_0;\ X_i \geq 0,\ i = 1, 2, \ldots, n \end{aligned}$$

where, $\boldsymbol{m}_0$ = *the target mean return.*

Problem 2 (Expected Return Maximization):

$$\begin{aligned} & minimize\ \boldsymbol{X}^T \boldsymbol{m} \\ & subject\ to, \\ & \sum_{i=1}^{n} X_i = 1 \\ & \boldsymbol{X}^T\ \boldsymbol{C} \boldsymbol{X} = \sigma_0^2;\ X_i \geq 0,\ i = 1, 2, \ldots, n \end{aligned}$$

where, σ_0^2 is the target return variance.

Ideally an investor would like to have both: *large return* and *small variance in the return*. This consideration leads to the following formulation.

Problem 3 (Joint Expected Return Maximization and Risk (Variance) Minimization):

$$\begin{aligned} &maximize\ \boldsymbol{X}^T\boldsymbol{m} - \lambda \boldsymbol{X}^T\boldsymbol{C}\boldsymbol{X} \\ &subject\ to,\ \sum_{i=1}^{n} X_i = 1;\ X_i \geq 0,\ i = 1, 2, \ldots, n \end{aligned}$$

where, λ is a non-negative number and is known as the *risk-tolerance parameter. The larger λ is, more is the weight placed on risk variation resulting in lower risk in the investment. On the other hand, the lower λ is, it will be more risky for the investor.*

Remarks: Problem 3 is more general than the other two. It was originally formulated by the Nobel Laureate *Harry Markowitz* (1952). Incidentally, it is worth mentioning that Professor Markowitz placed his Nobel prize in UC San Diego Library, where he was an Adjunct Professor.

Methodology

Many solvers exist to solve convex optimization problems. We have used CVXPY (https://www.cvxpy.org/ (Accessed on 31 July, 2020)) to solve portfolio optimization problem. We have taken the dataset from the lecture notes of Prof. Shabbir Ahmed.[1] The data contains closing stock prices of three company stocks (*IBM*, *Walmart*, and *Southern Electric*) on the last trading day of the month from November 2000 to November 2001. Using closing stock prices, we first calculate the rate of return for each stock across different months. The rate of return values are then used to compute the covariance matrix. Using these values, optimization problems of Sect. 5.3 are solved. The results of the most general case (Problem 3) are presented next.

Results

Maximum mean return ($\boldsymbol{X}^T\boldsymbol{m}$) with respect to different values of λ is plotted in Fig. 5. The optimum fraction of investment ($\boldsymbol{X}$) is first obtained by solving Problem 3. The fraction of investment of total money in each of the stocks to achieve optimum objective value is plotted in Fig. 6.

Observations: Based on the above results, the following recommendations can be made: An investor who wants to take more risk and at the same time likes to maximize his (her) return is advised to put nearly all of his (her) investment in Asset 3. Whereas, a less risky investor is advised to put most of the investment in Asset 2, and the remaining in Asset 1.

---

[1] https://www2.isye.gatech.edu/~sahmed/isye6669/notes/portfolio (Accessed on 31 July, 2020).

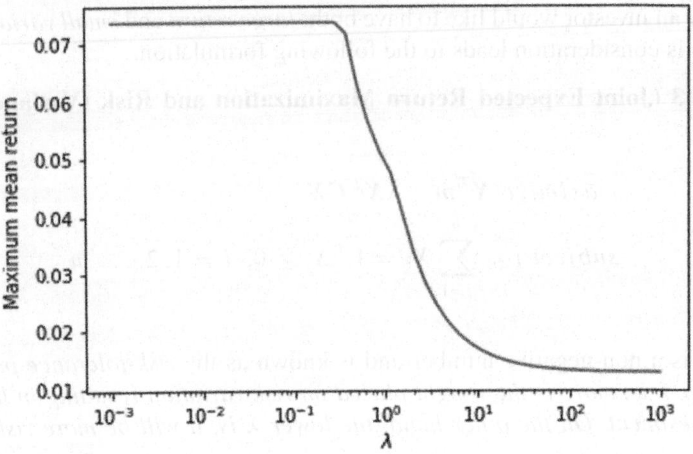

Fig. 5 Maximum mean return with respect to λ. *Source* Created by the authors

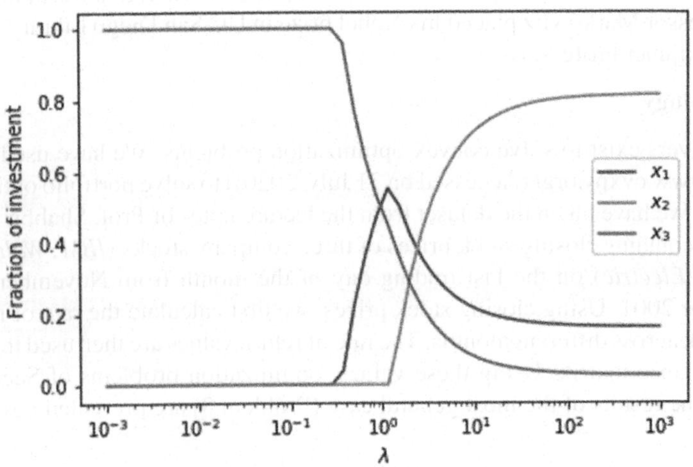

Fig. 6 Fraction of investment in each stock. *Source* Created by the authors

6 Concluding Remarks

In this paper, we have given a very brief introduction to machine learning with special attention to regression techniques for supervised learning. Some computationally effective algorithms, both for linear and nonlinear regression, in optimization setting, have been described. Conditions for optimality and some basic optimization algorithms, necessary to understand the optimization-based regression algorithms, have been briefly stated. Unfortunately, due to page limitations and considering the scope of this paper, Classification, another important machine learning technique, has not been considered in this paper.

Having said that, let us point out that the basic regression and classification techniques often are not capable of handling large and complex data arising in many applications. In recent years, deep Learning, another powerful machine learning technique, based on artificial neural networks, has been able to handle many such problems and has been successfully applied to important tasks, such as *speech and face and image recognition, social network filtering, video games* and many others. Several good books are available now on this subject, including the book by Goodfellow et al. (2016), and the more linear algebra-oriented book by Strang (2019).

Acknowledgements The author, Biswa Nath Datta, gratefully acknowledges the inspiring and motivating discussions he had with George Mathew, a freelance stock trader, on the "Portfolio Optimization Problem". This discussion led to the inclusion of the problem in this paper.

References

Bishop, C. M. (2006). *Pattern recognition and machine learning*. Springer.

Creswell, A., White, T., Dumoulin, V., Arulkumaran, K., Sengupta, B., & Bharath, A. A. (2018). Generative adversarial networks: An overview. *IEEE Signal Processing Magazine, 35*(1), 53–65.

Datta, B. N. (2010). *Numerical linear algebra and applications* (2nd ed.). Society for Industrial and Applied Mathematics.

Datta, B. N. (2020). *Numerical Methods and Analysis for Scientists and Engineers*. (Under preparation).

Freedman, D. (2009). *Statistical models: Theory and practice*. Cambridge University Press.

GitHub. (2020). https://github.com/biswajitsahoo1111/machine_learning_regression_and_optimization. Retrieved 31 July 2020.

Goodfellow, I., Bengio, Y., & Courville, A. (2016). *Deep learning*. The MIT Press.

Heath, M. T. (2018). *Scientific computing: An introductory survey* (2nd ed.). Society for Industrial and Applied Mathematics.

Krizhevsky, A., Sutskever, I., & Hinton, G. E. (2012). Imagenet classification with deep convolutional neural networks. *Advances in Neural Information Processing Systems* (pp. 1097–1105).

Montgomery, D. C., Peck, E. A., & Geoffrey, V. G. (2003). *Introduction to linear regression analysis*. John Wiley & Sons.

Nocedal, J., & Wright, S. J. (2006). *Numerical optimization* (2nd ed.). Springer.

Rao, C. R., & Mitra, S. K. (1971). *Generalized inverse of matrices and its applications*. Wiley.

Raschka, S., Patterson, J., & Nolet, C. (2020). Machine learning in python: Main developments and technology trends in data science, machine learning, and artificial intelligence. *Information, 11*(4), 193.

Rencher, A. C., & Schaalje, G. B. (2008). *Linear models in statistics* (2nd ed.). Wiley-Interscience.

Strang, G. (2019). *Linear algebra and learning from data*. Wellesley-Cambridge Press.

Watt, J., Borhani, R., & Katsaggelos, A. K. (2020). *Machine learning refined: Foundations, algorithms, and applications* (2nd ed.). Cambridge University Press.

The manufacturer's authorised representative in the EU is Springer Nature Customer Service Centre GmbH, Europaplatz 3, 69115 Heidelberg, Germany. If you have any concerns regarding our products, please contact ProductSafety@springernature.com

Printed and bound by CPI Group (UK) Ltd, Croydon, CR0 4YY
30/09/2026
02225296-0002